Butterworths Technical and
Scientific Checkbooks

Construction Site Personnel 4 Checkbook

J M Patchett
MSc, MIOB

Butterworths
London Boston Durban Singapore Sydney Toronto Wellington

First published 1983

© Butterworth & Co (Publishers) Ltd 1983

British Library Cataloguing in Publication Data

Patchett, J. M.
    Construction site personnel 4 checkbook.—
    (Butterworths technical and scientific checkbooks)
    1. Building sites—Management
    I. Title
    624'.068   TH438

    ISBN 0-408-00688-9
    ISBN 0-408-00685-4 Pbk

Typeset by Scribe Design, Gillingham, Kent
Printed by Thomson Litho Ltd, East Kilbride, Scotland

Contents

Preface

Site Studies (Personnel) is a TEC level 4 unit which forms an important part of many programmes for the awards of higher certificate and higher diploma within the construction sector. By studying the text and completing the questions at the end of each chapter, it is intended to ensure that the reader is able to meet all the objectives listed within the unit's guidelines.

The first nine chapters of the book relate to the first nine topic areas of the unit. They are, in the main, self-contained allowing the content to be studied in whichever order is preferred. The tenth and final topic area on 'Communication' has not been isolated but has been introduced within those chapters covering the subject matter of each objective.

Whilst the book endeavours to concentrate on relatively enduring principles and procedures it is necessary to refer to detail which is subject to change. The reader must be constantly alert, particularly where indicated in the text, to ensure that reference is made to current editions, legislation, codes of practice and regulations.

Finally I would like to thank all the many people who have provided assistance during the production of this book, and wish success to those who use it in the course of their studies and careers.

J M Patchett
Preston Polytechnic

Note to Reader

As textbooks become more expensive, authors are often asked to reduce the number of worked and unworked problems, examples and case studies. This may reduce costs, but it can be at the expense of practical work which gives point to the theory.

Checkbooks if anything lean the other way. They let problem-solving establish and exemplify the theory contained in technician syllabuses. The Checkbook reader can gain *real* understanding through seeing problems solved and through solving problems himself.

Checkbooks do not supplant fuller textbooks, but rather supplement them with an alternative emphasis and an ample provision of worked and unworked problems. The brief outline of essential data—definitions, formulae, laws, regulations, codes of practice, standards, conventions, procedures, etc—will be a useful introduction to a course and a valuable aid to revision. Short-answer and multi-choice problems are a valuable feature of many Checkbooks, together with conventional problems and answers.

Checkbook authors are carefully selected. Most are experienced and successful technical writers; all are experts in their own subjects; but a more important qualification still is their ability to demonstrate and teach the solution of problems in their particular branch of technology, mathematics or science.

Authors, General Editors and Publishers are partners in this major low-priced series whose essence is captured by the Checkbook symbol of a question or problem 'checked' by a tick for correct solution.

Butterworths Technical and Scientific Checkbooks

General Editor for Building, Civil Engineering, Surveying and Architectural titles:
Colin R. Bassett, lately of Guildford County College of Technology.

General Editors for Science, Engineering and Mathematics titles:
J.O. Bird and A.J.C. May, Highbury College of Technology, Portsmouth.

A comprehensive range of Checkbooks will be available to cover the major syllabus areas of the TEC, SCOTEC and similar examining authorities. A comprehensive list is given below and classified according to levels.

Level 1 (Red covers)
Mathematics
Physical Science
Physics
Construction Drawing
Construction Technology
Microelectronic Systems
Engineering Drawing
Workshop Processes & Materials

Level 2 (Blue covers)
Mathematics
Chemistry
Physics
Building Science and Materials
Construction Technology
Electrical & Electronic Applications
Electrical & Electronic Principles
Electronics
Microelectronic Systems
Engineering Drawing
Engineering Science
Manufacturing Technology
Digital Techniques
Motor Vehicle Science

Level 3 (Yellow covers)
Mathematics
Chemistry
Building Measurement
Construction Technology
Environmental Science
Electrical Principles
Electronics
Microelectronic Systems
Electrical Science
Mechanical Science
Engineering Mathematics & Science
Engineering Science
Engineering Design
Manufacturing Technology
Motor Vehicle Science
Light Current Applications

Level 4 (Green covers)
Mathematics
Building Law
Building Services & Equipment
Construction Technology
Construction Site Studies
Concrete Technology
Economics for the Construction Industry
Geotechnics
Engineering Instrumentation & Control

Level 5
Building Services & Equipment
Construction Technology
Manufacturing Technology

1 The construction team: roles, responsibilities and relationships

| 1.1 The construction process | 1.1.1 Evaluation
1.1.2 Design
1.1.3 Construction
1.1.4 Delivery |

| 1.2 Roles and responsibilities – building design and administration | 1.2.1 The client
1.2.2 The architect
1.2.3 The private quantity surveyor
1.2.4 The structural engineer
1.2.5 The services engineer
1.2.6 The building control officer |

| 1.3 Roles and responsibilities – building construction | 1.3.1 The main contractor's site staff
1.3.2 The sub contractor's site staff
1.3.3 The clerk of works
1.3.4 The factory inspector
1.3.5 Organisation for rehabilitation and renovation |

| 1.4 Roles and responsibilities – civil engineering | 1.4.1 The consultant civil engineer
1.4.2 The resident engineer
1.4.3 The main contractor's site staff |

| 1.5 Relationships between the construction team | 1.5.1 Formal relationships
1.5.2 In-Company formal relationships
1.5.3 Informal relationships |

| 1.6 Developments in organisation | 1.6.1 The client
1.6.2 The architect
1.6.3 The building main contractor |

1.1 THE CONSTRUCTION PROCESS

Construction is an industry which offers a wide variety of services providing structures to meet the needs of both individual and community. Each structure will differ in one or more of the following ways:

Purpose   :   Why?   Form   :   What?
Surroundings   :   Where?   Technology   :   How?

Efficient production will demand the exercise of many skills which are brought together in temporary teams; forming, disbanding and re-forming from project to project. Despite the differences each project passes through four identifiable stages:

Stage 1 Evaluation
Stage 2 Design
Stage 3 Construction
Stage 4 Delivery

Fig 1.1 shows that generally each stage will overlap; the degree of overlapping being dependent upon the form of contracting adopted.

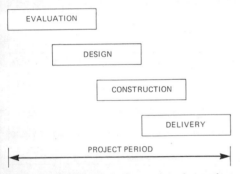

Fig 1.1 The four stages of a construction project

Up to the 18th century there was no separation, the designer/contractor and the client making decisions as the structure proceeded. However, as the complexity of building forms increased, the materials and technology became more demanding, and the two parallel disciplines of design and construction evolved, reinforcing the separation of the phases.

The organisation developed in sections 1.2 and 1.3 reflects this division which is still the dominant system within most sectors of the construction industry. There are, however, alternative methods of contracting which attempt to integrate the design and construction stages and some of these are examined in section 1.6.

1.1.1 PHASE ONE – EVALUATION

When an individual or group decide to build, having examined the possible utilisation of present buildings and renting or buying existing, it will be the result of a feasibility study initiated by the client but which will increasingly involve one or more of the construction professionals. The evaluation stage comprises:

Recognition of need;
Examination of alternatives;
Decision to build;
Appointment of construction professional;
Preparation of brief;
Appointment of other members of the building team;

Identification and inspection of site;
Consideration of finance, taxation and legislation;
Preparation of feasibility report;
Application for outline planning approval.

1.1.2 PHASE TWO – DESIGN

Phase two develops alternative approaches to the design, with cost, structural and legal implications being considered by the various members of the design team. A scheme detailing appearance, method of construction, outline specification and cost plan is then prepared for the client's approval, on receipt of which the design is finalised and production information prepared. The design stage comprises:

Consideration of alternative proposals;
Selection of scheme;
Application for full planning approval;
Completion of design and calculations;
Application for building regulations approval;
Preparation of design and construction drawings, schedules and specifications;
Preparation of bills of quantities, tender documents and contract particulars;
Seeking of tenders;
Analysis of response;
Appointment of contractor.

1.1.3 PHASE THREE – CONSTRUCTION

With a site, legal approvals, the selection of a contractor and the supply of full information, construction planning may commence. The contractor will ascertain his needs for production, managerial and administrative skills, for materials and for plant, all of which he may employ directly or through the use of sub-contractors.

When the construction works commence careful co-ordination of all such resources in relation to cost, quality and time is required.

The construction phase comprises:

Examination of all data;
Preparation of site layout;
Preparation of method statement, schedules and overall plan;
Application for necessary licences, insurances and permits;
Communication established with design team;
Preparation of short term plan;
Commencement of work on site;
Monitoring of production;
Administration of additional instructions;

Constitution of site meetings;
Negotiation of interim payments.

1.1.4 PHASE FOUR – DELIVERY

When the project, or a part of the project, is to all intents and purposes complete, the building will be accepted by the client following a formal inspection and handover. After the satisfaction of the defects liability the final account for the project is agreed and settled.

A review by all parties should then follow in which project performance is evaluated with the results providing feedback as a basis for future planning. The delivery phase comprises:

Inspection of work;
Completion of remedial work;
Organisation of insurance;
Provision of as-built drawings;
Provision of plant test certificates and instruction manuals;
Preparation of maintenance schedules;
Preparation of final account;
Compilation of analyses of building performance.

1.2 ROLES AND RESPONSIBILITIES : BUILDING DESIGN AND ADMINISTRATION

1.2.1 THE CLIENT

The client is the individual or group of individuals to whom the construction industry offers its services and the satisfaction of the client's requirements must be the highest priority. Many classifications are possible and the following form the major divisions:

Type of work
New work
   Replacement
   Additions
Work on existing buildings
   Maintenance
   Renovation
   Rehabilitation
Function of building (e.g. Residential, Administrative commercial, Health and Welfare)
Nature of employment of persons carrying out work
   Private sector
   Public sector
Nature of finance
   Private sector
   Public sector

The client's major responsibilities are to:
   Select the method of contracting;

Select the personnel;
Prepare brief and approve scheme;
Monitor progress;
Pay for work carried out.

1.2.2 THE ARCHITECT

Traditionally the architect has been the first building professional approached by the client at the earliest stages of the project and the role has, therefore, evolved with responsibilities for both design and administration.

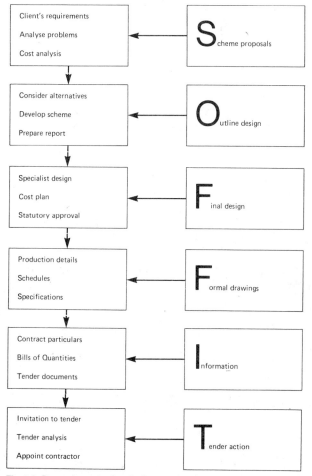

Fig 1.2 Procedures in the design stage of a traditionally organised construction project

The design function Within this function the true requirements of the client must be ascertained and translated into a building form which is acceptable to the client, whilst also satisfying the legislative requirements. It comprises an inter-related and complex set of procedures summarised in *Fig 1.2.*

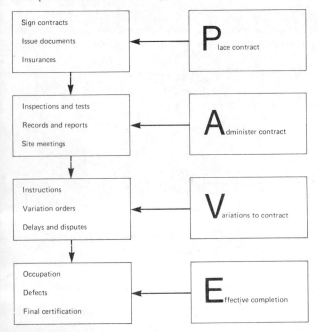

Fig 1.3 Procedures in the administration phase of a traditionally organised construction project

The administration function The architect's traditional role as leader of the building team carries responsibilities for the co-ordination and oversight of the activities of the other members of the building team. *Fig 1.3* summarises these procedures. The major responsibilities are to:

Offer competent advice on the appointment of specialists and selection of a contractor:
Advise on statutory requirements;
Oversee and co-ordinate the building team;
Issue all necessary information;
Supervise the work;
Administer the contract and finalise the account.

1.2.3 THE PRIVATE QUANTITY SURVEYOR

The private quantity surveyor advises throughout the project on financial and contractual matters and is appointed by the client but generally on the recommendation of the architect. If these responsibilities are to be carried out effectively the quantity surveyor should be appointed at the earliest possible moment and contribute fully to the design process. The major responsibilities are to:

Prepare preliminary costs plans;
Predict cost implications of design decisions;
Prepare budget to advise client of financial demands;
Prepare tender documents;
Agree equitable stage payments;
Monitor construction costs and variations in contract;
Agree equitable final account;
Analyse costs providing feed-back data.

1.2.4 THE STRUCTURAL ENGINEER

The structural engineer is similarly appointed by the client on the advice of the architect and is responsible for all those matters which affect the stability of the building structure. Being conversant with the mechanical strength of materials, the structural engineer will ascertain the size and location of elements in the foundations, walls, frame, floors and roofs. The major responsibilities are to:

Supervise investigations into site subsoil conditions;
Advise on possible structural forms, materials and approximate sizes;
Prepare and approve working drawings containing full details of structure and reinforced concrete bar bending schedules;
Approve structural details provided by other members of the building team;
Supervise erection of structural elements.

1.2.5 THE SERVICES ENGINEER

The provision within the building of a comfortable and efficient environmental control system may require expertise in one or more of the following areas:

Acoustics;
Air conditioning;
Heating;
Lighting;
Plumbing;
Sanitation;
Telecommunications;
Transportation;
Ventilation.

As the provision of such services requires co-ordination of one with the other as well as with the structure, the appointment of the respective specialist engineer at the earliest time is important. The major responsibilities are to:

Investigate and advise on services at design stage;
Prepare and approve working drawings and schedules related to services;
Supervise services within structure;
Consult external authorities regarding acceptability of design;

1.2.6 THE BUILDING CONTROL OFFICER

The building control officer is a representative of the local authority who checks to ensure that a proposed building meets the requirements of all appropriate legislation.

Work on site must not commence until satisfactory authorisation has been received in respect of both Building Regulations and planning permission.

If the drawings are considered satisfactory approval will be received with a series of inspection notices generally in the form of pre-printed post cards. The contractor must complete and return the cards at the following stages of the works when a building control officer will visit the site:

Commencement of works;
Excavation of foundations;
Concrete in foundations;
Damp-proof course;
Oversite filling;
Drain test before and after covering;
Completion and occupation of building;

1.3 ROLES AND RESPONSIBILITIES : BUILDING CONSTRUCTION

1.3.1 THE MAIN CONTRACTOR'S SITE STAFF

The organisation structure of a contractor is designed to meet the set of problems posed by the technical and physical complexity of a building in addition to reflecting company policy generally. The roles and relationships will therefore vary to a large degree from company to company and even from site to site within the same company.

The Site Manager, as senior representative on site, is generally responsible to an office based contracts manager. His function is the supervision and control of the execution of the works ensuring completion within the planned cost and time to the quality required under the terms of the contract. The major responsibilities are to:

Advise and assist in the overall planning;
Plan and coordinate the supply of resources;
Monitor and control progress and standards;
Attend site meetings and liaise with the client's representative;
Provide feedback as required and conduct all aspects of the works in accordance with company policy;
Act within existing legislation in respect of records, safe methods of working and the employment of operatives.

The General Foreman will be subordinate to the site manager on a larger contract but may, on smaller sites, be the highest on-site authority. This will not necessarily involve any additional responsibility as in the latter case closer supervision will be exercised by a head office based manager.

The work involves the transmission of instructions to the trade foremen and general supervision of the rate and quality of output. General foremen are almost invariably craft trained and this experience assists communication and provides a basis upon which to solve the problems which arise during the course of the works. The major responsibilities are to:

Coordinate and control the work of the trade foremen;
Ensure the efficient and safe utilisation of resources;
Advise site management on practical constructional problems;
Monitor and report on industrial relations;
Liaise with the clerk of works.

8

The Trade Foreman issues instructions to and supervises the work of a particular trade from whose ranks he is invariably selected. The major responsibilities are to:

Ensure proper instructions are received by operatives;

Monitor progress and report on resource requirements;

Monitor the quality of work;

Liaise with other trades foremen to ensure operations are coordinated.

The Ganger has similar duties to the trade foreman, having responsibilities as above, but in the supervision of semi and unskilled operatives.

The Operative carries out the construction work within the building process. Skills may be gained through an apprenticeship leading to craftsman status; through short training courses or simply through experience. In any case the operative should be capable of carrying out allocated tasks in a reliable and efficient manner. The major responsibilities are to:

Carry out alloted tasks to the required standard and at an acceptable rate;

Observe safe methods of working.

The Site Engineer is responsible to management for the accurate setting out of the works and components of the structure in addition to quality control of materials. The major responsibilities are to:

Set out the grid lines and the structure;

Establish levels as required;

Participate in the quality control system;

Provide feedback information.

Support staff To provide a complete range of services to site management, many organisations employ additional specialists who, dependent upon the size of the site, may be based directly on site and be responsble to the site manager. When the size does not justify this the support staff may be based at head office under the contracts or a functional manager and contribute to more than one site. The following is a list of such staff:

Bonus surveyor;

Builders' quantity surveyor;

Buyer;

Clerical staff;

Planner.

1.3.2 THE SUB-CONTRACTOR'S SITE STAFF

The Contracts Supervisor Generally a sub-contractor will have a small number of men on each site and the contracts supervisor will travel from site to ensure the efficient execution of the works, assisted by a number of support staff who will invariably be office based. The major responsibilities are to:

Liaise with the site manager in regard to attendance;

Coordinate the supply of all resources;

Attend site meetings as required;

Ensure the quality and safety of the works.

The Chargehand On site, each sub-contractor must have one person in charge. If the site is large then that individual may be in full time supervision, but more often the chargehand will be a working supervisor. It is important that he and the general foreman are aware of the limits of authority which will generally be restricted to day to day organisation only.

The Operative Operatives working for sub-contractors have skills gained on the same bases as those of the main contractor. They are generally on site for much shorter periods of time and tend to travel further afield working on a greater variety of contracts.

1.3.3 THE CLERK OF WORKS

The clerk of works is the client's representative who may be resident on larger sites, or be a regular visitor, particularly at critical phases of a small site. The clerk of works liaises closely with the architect and will therefore generally be appointed on his nomination. The primary duty is to ensure that the constructed works conform to the specifications laid down in the contract documents. The clerk of works has authority over the main contractor only through the architect and must exercise that authority with persuasion and tact rather than dominance.

The major responsibilities are to:

Inspect materials and workmanship;

Arrange suitable tests;

Maintain written records as requested by the architect;

Advise the architect of any informational requirements;

Endorse the main contractor's record of labour and materials expended in work in variance of the contract.

1.3.4 THE FACTORY INSPECTOR

During the course of the works it is the contractor's duty to ensure compliance with all legislative requirements relative to the safety, health and welfare of all employees and visitors to site. The factory inspector, employed by H.M. Factory Inspectorate, has powers to issue instructions for the rectification of deficiencies in a given period of time, or prohibit the carrying out of further work until serious contraventions have been remedied. The factory inspector has powers to enter premises, take samples for examination, require persons to give information and to inspect statutory registers.

1.3.5 ORGANISATION FOR REHABILITATION AND RENOVATION

Rehabilitation and renovation of vacant property on a large scale may be carried out generally with the organisation structure previously described, but small scale works and that to occupied premises create different demands. There are a large number of craftsmen proprietors working as sole traders or within partnerships whose market comprises solely rehabilitation and renovation. Much of this business is gained through personal recommendation and therefore good workmanship, reliability and effective customer relations are essential prerequisites of long term success.

Large scale contracts within occupied premises, the refurbishment of local authority housing for example, require special skills for which alternative procedures have been developed (illustrated in *Fig 1.4*), reflecting the particular needs of the situation.

The Site Manager Modernisation of people's homes particularly whilst they are in residence creates problems in regard to communication with occupants, scheduling of

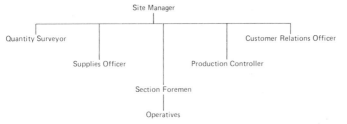

Fig 1.4 On-site organisation for a large rehabilitation and renovation project

work, material control and personnel selection. The site manager has first to establish the extent of the work and this is done through a pre-entry inspection with the Local Authority. Agreement can also be reached on existing damage which otherwise may be claimed as contractor's liability.

The Quantity Surveyor Whilst the general extent of the work will be standardised, different house types and the pre-entry inspection will inevitably lead to individual differences between the work on each dwelling. The aquisition of an architect's instruction and the preparation of a full and fair claim is the responsibility of the site based quantity surveyor.

The Supplies Officer To take advantage of bulk buying, yet to issue materials on a house to house basis creates problems of supplies control. The supplies officer is responsible for the efficient ordering, receipt, storage and issue of all necessary materials and plant.

The Production Controller To prevent inconvenience to the house-holder the accurate scheduling of starts and completions is essential, this being the primary responsibility of the production controller.

The Customer Relations Officer The responsibilities of this position are related to the communication needs of contractor customer relations. Initially general meetings can explain the approach to and extent of the work, and regular visits before, during and after the work will ensure minimum disturbance and stress.

The Section Foreman Dependent upon the size and handover rate of the contract, it may become necessary for general foremanship to be shared either through stage specialisation or through house numbers. Additional demands are placed on the foreman because of the high degree of contact with the householder.

The Operative It is not sufficient that the operative be a good craftsman, needed also are consistency, dependability and the ability to earn the trust of the occupier.

1.4 ROLES AND RESPONSIBILITIES: CIVIL ENGINEERING

There is no clear distinction between building and civil engineering, the two major areas of the construction industry, most contracts containing features of both. As a

11

generalisation however the building industry is concerned with projects which enclose space within which man undertakes his various activities; housing, schools, factories, hospitals and churches for example. Civil engineering on the other hand, provides the infrastructure which services man's convenience; roads, power stations, sewage installations, bridges and harbours for instance.

As many companies carry out both types of work the roles of the two organisations are similar but differentiated by the scale of the project, the fewer and more repetitive production operations and the smaller range of construction materials. Whilst the skills of the contractor are similar in both sectors, the design and administration have produced a divergence of personnel.

1.4.1 THE CONSULTANT CIVIL ENGINEER

The consultant civil engineer acts as the overall project manager in a similar way to the architect, but has greater powers in that other design specialists are directly appointed. The major responsibilities are to:

Assist the client in the preparation of the brief;

Appoint specialists as necessary;

Undertake the site investigation;

Design the works, preparing models and data as required;

Advise on contractor selection and prepare documentation;

Implement the contract and oversee quality and production rate;

Administer variations and additions, and make stage payments;

Oversee the maintenance period and agree the final account.

1.4.2 THE RESIDENT ENGINEER

As a result of the size and complexity of the works a resident engineer is usually appointed who is based on site and constantly available to represent the client's interests through the consultant civil engineer. The major responsibilities are to:

Oversee the work for line, level, materials and workmanship;

Execute tests on site and keep a diary of happenings;

Measure quantities of work in agreement with contractor's staff;

Record the facts of any work for which the contractor may demand extra payment;

Check and record the progress of work and full details of deviations.

1.4.3 THE CONTRACTOR'S SITE STAFF

As indicated in 1.3.1. in respect of the building main contractor the on-site organisation will vary in accordance with company policy and physical complexity of the site. *Fig 1.5* however shows an example of a contractor's organisation designed to meet the problems posed by the larger site of a more complex nature.

The Agent is the experienced engineer who acts as chief executive within the site organisation. The role is to see that the works are administered effectively in accordance with the contract documentation and to the requirements of the resident engineer.

The Sub-agents have direct control of part of the works, exercising authority over the production through section engineers and general foremen.

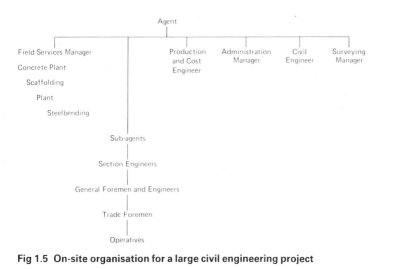

Fig 1.5 On-site organisation for a large civil engineering project

The General Foreman supervises the day-to-day disposition of labour and plant and the flow of materials.

Service personnel The agent may be supported by one or more of the following service departments dependent upon the degree of decentralisation of control and size of site.
Field Service Manager – responsible for the control of service departments.
Production Control and Costing Engineer – responsible for the planning, progress and costing functions.
Administration Manager – responsible for the efficiency of non-technical administration.
Civil Engineer – responsible for the accuracy of the works.
Managing Surveyor – responsible for the calculation of quantities for payment purposes.

1.5 RELATIONSHIPS BETWEEN THE CONSTRUCTION TEAM

Given that the satisfactory execution of a construction project requires the contribution of many and varied participants, their inter-relationships are critical. Such relationships are the resultant of many factors both formal and informal which include the role concept and the contractual implications in addition to individual personality.

1.5.1 EXTERNAL FORMAL RELATIONSHIPS

The construction process creates a great number and a varied range of contracts. The law of contract establishes the rights and liabilities of parties, differentiating contracts and agreements. The rights and obligations of parties must be expressed clearly in writing and not be left to the subjective interpretation of an oral contract. A number of standard agreements have been developed containing clauses to suit a variety of contingencies. The advantage of such agreements is that parties become familiar with their operation and legal precedent clarifies points of contention.

JCT Standard Forms In 1980 the Joint Council Tribunal, a body representing the professional institutions and employers' associations, revised the following forms:
JCT Standard Form.
Local authorities with quantities
Local authorities without quantities
Private with quantities
Private without quantities
Local authorities with approximate quantities
Private with approximate quantities
Form of tender by nominated supplier.
Agreement for minor building work.
Design and build form.
Also introduced were forms NSC/1,2,3,4 and 4a which relate to the engagement of nominated sub-contractors.

NFBTE The National Federation of Building Trades Employers publish two forms of contract:
NFBTE Standard form of sub-contract.
NFBTE Form of contract for use when the contractor is to design and build.

Government contracts All building and civil engineering works carried out for government departments are subject to the provisions of the following:
GC/Wks — Form of general conditions for building and civil engineering works.
GC/Wks/1 — Major government works.
GC/Wks/2 — Minor government works.

Institution of Civil Engineers Civil engineering work not covered by the above generally uses the Form of general conditions of contract for works of civil engineering construction.

1.5.2 INTERNAL FORMAL RELATIONSHIPS

Within the organisation of any company or practice, the allocation of responsibilities

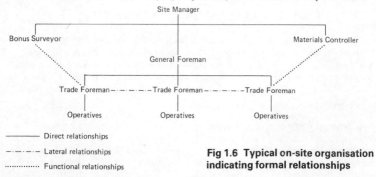

Fig 1.6 Typical on-site organisation indicating formal relationships

creates a formal network of inter-relationships illustrated in *Fig 1.6*. The major classifications are:

Direct; Lateral; Functional.

Direct relationships exist between superior and subordinate and involve the right to issue instructions and the obligation to carry them out.

Lateral relationships exist between two subordinates having equal levels of authority and being answerable to the same superior. This relationship involves a co-operative responsibility.

Functional relationships exist between service departments and production management and carry with them the responsibilities of expertise. It is important to recognise that employees should clearly understand the relationships and responsibilities inherent in their positions.

1.5.3 INFORMAL RELATIONSHIPS

It is neither possible nor desirable to control all activities which make up the organisation of the construction industry. It has been suggested that people will amplify and modify the system to suit their needs which may not necessarily match the needs of the organisation. The informal organisation thus created can therefore be a constructive or a disruptive force.

It is the informal organisation that provides natural leaders who are not necessarily recognised by the formal and the 'grapevine' of information flow will flourish particularly where confidential information is concerned.

It has, however, also been suggested that the existence of the informal organisation simply reflects deficiencies within an organisation which does not provide the opportunity for the participants to satisfy their status and social needs. If employees are given defined areas of responsibility and freedom of decision making within those areas then the formal organisations will absorb the informal each complementing the other.

1.6. DEVELOPMENTS IN ORGANISATION

Inspired by government reports and commercial reorganisation some changes have occurred in the recent past which have led to a redefinition of the traditional roles and responsibilities previously described. This section attempts to summarise some of those changes.

1.6.1 THE CLIENT

The traditional approach to contracting considers that the client has little to offer by way of skill and knowledge to the construction process, and for some this is so. However many organisations, having a continuous need for certain skills find it more efficient to directly employ their own staff. Local authorities for example may employ architects, quantity surveyors and engineers; and some, having direct works departments, act as completely autonomous units.

1.6.2 THE PROJECT MANAGER

The client may employ, or appoint on a professional basis, an individual whose responsibilities are to co-ordinate and control the project from evaluation to delivery.

This service has come to be known as project management. The role has to achieve integration of all functions through the clear definition of objectives. Management skills are of utmost importance and whilst the project manager will come generally from the ranks of the established construction professionals and thus understand and respect the various disciplines it is generally undesirable to fulfil any individual function, acting solely as a leader and co-ordinator.

1.6.3 THE ARCHITECT

Whilst some architects, released by certain contracting methods from their administrative role, have enjoyed a concentration on the design function, others have centralised the organisation of the project team. Larger practices have done so by employing not only architects but planners; civil, structural and service engineers; quantity surveyors and interior designers. This has of course necessitated an advanced clerical and technical support facility.

Many such practices organise themselves into project groups comprising the necessary expertise for each contract and headed by a project leader who represents the team in all major dealings with the client.

1.6.4 THE BUILDING CONTRACTOR

The need for greater liaison between design and construction stages of a project has led to changes in contractor organisation with particular reference to the services offered directly to the client.

Negotiated Contracts The elimination of the competitive element in the gaining of contracts places additional responsibilities on the client and contractor in relation to the establishment of prices and standards. Negotiation does permit a greater overlap between design and construction resulting in closer integration and generally time savings.

There are many methods by which the final charge for a building may be computed but perhaps the most common method is the 'cost plus fee' system whereby the prime cost of the works is paid to the builder along with a management fee to include overheads, supervision and profit. The fee may be a percentage of the prime cost or a fixed fee, which provides an incentive for the contractor to operate as effectively as possible.

Design and Build The complete absorption of the design sphere into construction results in a form of contract generally referred to as a design and build service. This service has been offered in the speculative housing field for some time but is becoming increasingly popular in industrial and commercial fields, particularly where designs are repetitive or use standard components.

Management Contracting Many projects involve the main contractor in little direct production, the majority being let on sub-contract basis. The natural progression is to relinquish all direct production work and concentrate solely upon co-ordination and control. The management contractor may become involved in the project from the outset committing himself to a percentage fee and yet still provide the competitive element through the selection of sub-contractors.

BIBLIOGRAPHY

1   The Aqua Group, *Contract administration for architects and quantity surveyors,* Granada, St Albans, 5th ed (1981)
2   The Aqua Group, *Pre-contract practice for architects and quantity surveyors,* Granada, St Albans, 6th ed (1980)
3   The Chartered Institute of Building. Annotated bibliographies.
    44 Role of the professional in the building team
    45 Client — Building team relationships
    49 Management contracting
    CIOB Ascot
4   The Chartered Institute of Building. *The management and organisation of building maintenance : a Review of the Literature 1970—1978,* CIOB, Ascot (1979)
5   The Chartered Institute of Building. *The practice of site management,* Vol 1 & 2, CIOB Ascot (1980)
6   The Chartered Institute of Building. *Occasional Paper 20 : Project Management in Building,* CIOB Ascot
7   GLC Department of Architecture and Civic Design. *Handbook for the Clerk of Works,* The Architectural Press, London 2nd ed (1979)
8   Higgin G, and Jessop N., *Communications in the building industry,* Tavistock Publication (1965)
9   The Institution of Civil Engineers. *Civil Engineering Procedure,* ICE London (1976)
10 Jones, G.P., *A new approach to the 1980 Standard Form of Building Contract,* Construction Press, Lancaster (1980)
11 Jones, G.P., *A new approach to the ICE Conditions of Contract,* Vol 1 & 2 Construction Press, Lancaster (1975)
12 Lee, R., *Building maintenance management,* Granada, St. Albans, 2nd ed (1981)
13 National Federation of Building Trade Employers, *Managing a smaller building firm,* NFBTE London (1980)

EXERCISES

RECALL QUESTIONS

1   Complete the four identifiable stages of a construction project
    (a) Stage 1   E . . . .   (c)   Stage 3   C . . . .
    (b) Stage 2   D . . . . .   (d)   Stage 4   D . . . .

2   Within traditional contract procedure, in which stage do the following activities occur
    (a) Site inspection   Stage . . .
    (b) Performance evaluation   Stage . . .
    (c) Method statement prepared Stage . . .
    (d) Planning approval sought   Stage . . .

3   Within traditional contract procedures indicate the member of the building team who carries out the following
    (a) Selects the method of contracting;
    (b) Advises on structural forms;
    (c) Sets out the gridlines of the structure;
    (d) Administers the form of contract;

(e) Checks that the building meets legislative requirements;

(f) Controls the work of the trade foremen;

(g) Plans and co-ordinates the use of resources on site;

(h) Prepares the financial budget.

4 Within traditional contracting procedure indicate the members of the civil engineering construction team who carry out the following

    (a) Manages all site operations;   (d) Tests materials on site;

    (b) Appoints specialists advisors;   (e) Calculates quantities of work;

    (c) Supervises the disposition of labour;   (f) Plans and progresses the work.

5 Identify the type of relationship that exists in an organisation between

    (a) Superior and subordinate   ....

    (b) Specialists and production management   ....

    (c) Equals with the same superior   ....

QUESTIONS REQUIRING SHORT ESSAY ANSWERS (15 – 20 minutes)

1 State the advantages and disadvantages of the use of a standard form of building contract.

2 Prepare a possible organisation chart for the site shown on drawing no. —— which is 200 kilometres from the main contractor's head office.

3 Describe briefly the major responsibilities of three of the following members of the construction team

    (a) Architect;

    (b) Building Control Officer;

    (c) Building or Civil Engineering General Foreman;

    (d) Resident Engineer;

    (e) Sub-Contract Contracts Supervisor;

4 Briefly describe the four stages of a construction project.

5 List the advantages to the client of the appointment of a project manager.

QUESTIONS REQUIRING LONGER ESSAY ANSWERS (30 – 45 minutes)

1 From the client's point of view compare the services offered by the companies engaged in the following

    (a) Local authority housing contracts gained through competitive tender;

    (b) Industrial building contracts gained through design and build agreements;

    (c) Commercial contracts gained through management contracting agreements.

2 Discuss the problems inherent in the appointment of functional specialists within a building contractor's organisation.

3 Describe the advantages and disadvantages which result from the specialisation of the various members of the construction team.

4 Compare the role and responsibilities of the person with ultimate on-site authority on the following contracts

    (a) Rehabilitation of local authority housing;

    (b) A large modern hospital complex;

    (c) Major road construction.

5 Discuss in detail the disadvantages to the client of the division between the design and construction stages of a building project.

18

2 Motivation of construction workers

2.1 Motivation theory	2.1.1 Scientific management 2.1.2 The Hawthorne investigations 2.1.3 Satisfaction theory 2.1.4 Motivation – Hygiene theory 2.1.5 Theory X and Theory Y 2.1.6 Expectancy theory 2.1.7 Equity theory 2.1.8 Reward systems
2.2 Direct financial systems	2.2.1 Daywork 2.2.2 Piecework 2.2.3 Hours saved systems 2.2.4 Geared hours saved systems 2.2.5 Group schemes 2.2.6 Plus rates 2.2.7 Job and finish 2.2.8 Profit sharing 2.2.9 Principles of a good scheme
2.3 Indirect financial systems	2.3.1 Security of earnings 2.3.2 Subsidised services 2.3.3 Domestic expenses 2.3.4 Other schemes
2.4 Non-financial systems	2.4.1 The physical working environment 2.4.2 The social working environment 2.4.3 The work itself
2.5 Financial reward systems in the construction industry	2.5.1 Direct financial systems 2.5.2 Indirect financial systems 2.5.3 Establishing a direct financial system
2.6 Non-financial reward systems in the construction industry	2.6.1 The physical working environment 2.6.2 Job security 2.6.3 Health and safety 2.6.4 The social working environment 2.6.5 The work itself

2.1 MOTIVATION THEORY

Management has been defined as the achievement of objectives through people. Organisational objectives can only be achieved through the voluntary participation of groups of workers. Management's task, therefore, is to maintain a high level of voluntary participation within the firm.

Motivation has many definitions even within management theory but most are concerned with the process of initiating and directing behaviour. The way in which any individual manager will seek to carry out this function will be dependent upon certain underlying assumptions of man. These have been classified as follows:

Rational Economic Man Man is a passive participant who is to be controlled and directed and is motivated by economic needs.

Social Man Man gains his identity through relationships with others and therefore through leadership and knowledge of group behaviour, successful social relationships must be established within the work situation.

Self Actualising Man Man is self motivated and self controlled and will, in the right conditions, seek responsibility and require no external control.

Complex Man Man has many and complex motives for his behaviour which vary with time and circumstance. His response to direction will vary with his evaluation of its appropriateness.

The questions 'Why does man work?' and 'Why does one man work harder than another?' have therefore many different answers of varied complexity and many theories attempt to provide solutions which can be applied generally throughout industry. The theories of work motivation have been classified in the following manner:

Incentive theories These theories suggest that man will direct his efforts to attain desired tangible rewards because:
1. He wants the reward;
2. He believes the reward is worth the effort;
3. He is in complete control of his rate of production.
The most common reward is money which can be exchanged for any desired available object or service.

Satisfaction theories The assumption here is that man is motivated by a job which allows him to meet a variety of needs. It requires assumptions regarding the relationship between job satisfaction and productivity. Whilst it is established that consideration of the needs of man generally leads to reduced labour turnover and a lower absentee rate the assumption that a satisfied worker is a more productive worker has yet to be proved. It has been suggested for instance that high productivity causes job satisfaction rather than the other way round.

Intrinsic theories Man's identification of, and search for, an ideal self image leads to a constant attempt to close the gap between what a man is and what he would like to be. The gap will tend not to close as the image is developed following each achievement. The rewards of industry have traditionally been based on the external needs of man but intrinsic theories seek to suggest that only through a redesign of task and relationships introducing involvement and participation will man secure the achievement of self improvement.

In America, in the early 1900s, a group of engineers and managers headed by
F.W. Taylor developed a system of organisation which they knew as Scientific
Management in an effort to improve output on the factory floor. Following scientific
analysis production operations were rationalised and simplified and the planning
process formalised. Control procedures were established to discover and correct
deviations to the plan.

This approach clearly separated the functions of managerial and production workers
and the latter were specially selected and trained in a narrow field of activity. For the
resultant increase in productivity through specially devised incentive schemes the
workers received 'a fair day's pay'.

Taylor's writings make it clear that his motives were not exploitative and that the
efficiency induced would result in both increased earnings for the production worker
and a cheaper product which would find a wider market. He was disappointed to find
that his methods of control were adopted without a similar adoption of the philosophy
of mutual advantage. Organised labour, fearing job security and exploitation, resisted
the introduction of 'Taylorism' resulting in conflict and an enquiry by the American
Government.

It is fair to say that much of the philosophy and assumptions of scientific management are expressed by today's industrial management. Systematic analysis of the
situation results in a carefully planned work schedule which is controlled through rigid
control procedures, none of which appears inherently wrong. However it has been
suggested that the main assumptions upon which this approach is based are that:
1 Workers are typically lazy, dull, aimless, and mercenary.
2 The only way to make a man work harder is through financial incentive.
3 All initiative must be removed from the worker.
4 Success is measured solely in terms of speed and cost.
5 The worker is secondary to the work.
6 All work can be evaluated and permanent, accurate standards set.

2.1.2 THE HAWTHORNE INVESTIGATIONS

In the 1920s at the Western Electric Company's plant at Hawthorne, Chicago, a classic
series of studies were devised. Following scientific method, all the known factors which
might affect performance were brought under control.

The first investigation established two groups of workers doing the same job, a
control group whose conditions were unchanged and an experimental group whose
lighting intensity varied, the intention being to establish the optimum lighting
conditions. It came as no surprise that an initial raising of the lighting level resulted in
increased production by the test group. What was surprising was that the production
of the control group also increased and subsequent lowering of the lighting level even
to that of moonlight failed to depress this rise in productivity.

It was clear that the experimental conditions had failed to control at least one
significant factor and between 1924 and 1932 under the direction of E. Mayo a series
of further investigations into work groups, reward systems, work patterns and supervision were carried out. In addition an interview programme was established with
20,000 employees participating in increasingly informal sessions to monitor attitudes.
From the results of the investigations the following generalisations were made:
1. All management has a human element.
2. Non-financial incentives are important.

3. The organisation of production is the starting point of management.
4. The style of supervision is important.
5. A manager/worker dialogue must establish communication between groups.
6. Symptoms must be distinguished from causes during the investigation of human behaviour.

There are, no doubt, omissions and simplifications in this early work in the social sciences but the establishment of the human factor within the work group established social psychology of industry as a valid and important area of study.

2.1.3 SATISFACTION THEORY

In 1943 A.H. Maslow first suggested on the basis of clinical psychological observation that the human is a wanting animal motivated by a desire to satisfy specific needs. He suggested that the needs are arranged sequentially and that the lower needs must be satisfied before higher needs emerge as motivators of behaviour, (*Fig 2.1*).

Fig 2.1 Maslow's need hierarchy

Physiological needs The basic bodily needs of, for example, food, drink, and shelter are considered to be the most potent if unsatisfied. Conscious social behaviour may, in times of extreme stress be overridden by the need to survive. Once the need is satisfied however it ceases to be a determinant of behaviour until such times that it returns.

Safety needs Society as an organisation exists to protect the individual from external threats to personal safety. The armed and police forces; pension schemes; savings and insurance policies are all manifestations of the protection sought by individuals in an ordered society.

Social needs After the satisfaction of physiological and safety needs that of belonging emerges. The individual seeks relationships with others, striving to obtain membership of groups within which he can both give and receive satisfaction.

Esteem needs The satisfaction of this need requires that the individual achieves a state of self confidence and adequacy based upon feelings of capability and achievement in addition to the respect of others.

Self-actualisation needs This need requires growth and development for satisfaction until the individual feels that he has achieved everything of which he is capable.

Working in the 1970s C. Alderfer suggested that the five needs can be simplified to three which he called, **Existence**, **Relatedness** and **Growth**. Existence relates to Maslow's physiological and safety needs; relatedness to belonging and part of esteem and growth is concerned with the desire to be creative and achieve one's potential within an organisation. The two theories differ most importantly in that Alderfer rejects the hierarchical nature of needs explaining that all three operate simultaneously and satisfaction in one area may act as substitute for another. Also it is suggested that their relative importance will vary between individuals and even within the same individual dependent upon the circumstances.

2.1.4 MOTIVATION – HYGIENE THEORY

Following a programme of practical research, F. Herzberg in 1959 suggested that those factors which, when present, lead to satisfaction are different from those factors which, when absent, lead to dissatisfaction.

The latter conditions he called hygiene factors and include working conditions, salary, relationships with others, company policy and supervision. The argument is that these factors may answer the question 'Why work here?', but to seek an answer to the question 'Why work hard?' a different set of factors must be considered.

These factors he called motivators which relate to areas previously called self actualisation and growth. They include advancement, responsibility, the work itself, recognition and achievement.

2.1.5 THEORY X AND THEORY Y

D. McGregor, an American academic, suggested in the 1950s that two sets of propositions could be listed relative to man in an industrial setting, and that managerial approach was based upon the acceptance of one set or the other.

Theory X established man as idle, lacking in ambition, resistant to change and to the needs of the organisation.

Theory Y, on the other hand, suggested that if this state exists it does so because of a reaction to faults in the environment and that if conditions are suitable man will behave responsibly and diligently seek out patterns of behaviour which will assist the organisation to meet its objectives.

2.1.6 EXPECTANCY THEORY

Expectancy, suggested in the 1960s by V.H. Vroom and E.E. Lawler argues at the basic level that effort will only be forthcoming if:
1. A relationship exists between effort and performance
2. A higher reward will follow higher performance
3. An attractive reward system is available
An individual's performance will depend upon his willingness to display effort, his possession of the requisite skills and abilities and his judgement of the methods he must employ to reach his goals.

Fig 2.2 Expectancy theory according to Vroom and Lawler

Fig 2.2 diagramatically represents this theory showing the feedback loop through which the individual will alter his beliefs in the light of his experience and hence result in an alteration of effort.

2.1.7 EQUITY THEORY

At approximately the same time J.S. Adams suggested that satisfaction and dissatisfaction relate to beliefs regarding the equity or fairness existing in the work situation. Equity is a state in which effort and reward are balanced when viewed in comparison with others both inside and outside the organisation.

The suggestion is that if injustice is considered to occur, various strategies are available to the individual:

1 Alter inputs — Reduce or increase quality or quantity of work or skills.
2. Alter outcomes — Take steps to increase or alter rewards.
3 Alter beliefs — Reassess the importance attached to inputs and outcomes.
4 Opt out — Leave job.
5. Change model — Compare with someone else.

2.1.8 REWARD SYSTEMS

The basic premise of most reward systems is that they result in an increase in productivity. There are however two major difficulties; firstly the definition and measurement of productivity and secondly the correct attribution of any increase to a particular element of the reward system.

Productivity can best be considered to be the degree of efficiency with which a resource is used, and could in labour terms be considered the relationship between the output and the size of the labour force although neither figure is easy to ascertain. This then increases the difficulty within the second element as productivity is influenced by much more than the effort of the labour force.

Within the production process; the design, the ease of building and management efficiency are major factors affecting productivity and therefore the assessment of the success or otherwise of the reward systems adopted by the construction industry is difficult.

Since the 1950s reward systems in the industry have been based to a great extent on the use of financial incentive schemes on the basis that:

1 Money is an attention getter;
2 Management is results orientated and believe that accomplishment should be rewarded
3 Money is more readily quantified than psychological factors.

If there is little direct evidence that financial incentive schemes result in increased productivity their contributions to the retention of labour cannot be denied. It has been suggested that there are four main reasons why people leave their jobs:

1 For pay and promotion;   3 For better job content;
2 For better work organisation;   4 For reasons of age.

Analysis has indicated that whilst a degree of turnover is considered essential for the generation of new ideas its excesses can be contained by the ability of the work situation to meet a peron's need for recognition, security, participation and autonomy.

Whether or not such systems lead to higher productivity in the longer term has been a question asked by many researchers but the answering suffers from the same problems as did the financial incentive scheme. Three basic questions are posed:

1 Does satisfaction lead to performance?

24

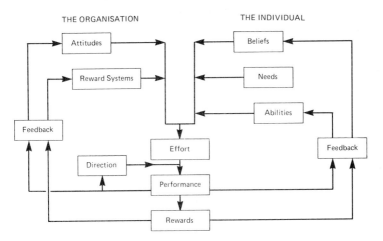

THE ORGANISATION   THE INDIVIDUAL

Fig 2.3 The process of motivation

2 Does performance lead to satisfaction?
3 Or are both affected by a larger and more variable set of conditions?

A link appears to have been established between satisfaction and performance although the evidence does not point to the exact nature of the relationship. It does however appear that rewards result in satisfaction which in turn results in reduced absenteeism and reduced labour turnover.

The reward of current performance requires that the manager recognises the individual needs by the establishment of a system which is capable of providing differential rewards based on both performance and need. (*Fig 2.3*).

2.2 DIRECT FINANCIAL SYSTEMS

The widely held traditional view that people work harder when they are offered a direct financial reward based on measured output has led to the development of many such schemes, the objectives of which are:
1 To increase productivity.
2 To encourage more efficient methods of working.
3 To provide increased earnings without increasing unit costs.

2.2.1 DAYWORK

Daywork provides for the payment of an hourly rate for attendance at work and is used for highly skilled work or for support operations like maintenance and inspection. There may be no relationship to productivity other than the expectations of management or, in measured daywork, a set target may have to be reached. In itself it does not differentially reward productivity and is often supplemented by some form of reward system.

25

Advantages:	Disadvantages:
1 Easy to calculate and understand.	1 No differential rewards.
2 Low level of wage computation.	2 Slow worker gains at the expense of the fast.
3 Facilitates flexibility of labour.	3 Supervision of attendance and production rate required.
	4 Production forecasts difficult.

2.2.2 PIECEWORK

Piecework provides a uniform payment of a price for each unit of work so that the reward is geared directly to the workers' production (*Fig 2.4*) The system is most often used where average job performance in a repetitive activity can be assessed. A variant expressed in terms of hours per item as opposed to a financial sum requires recalculation of targets following wage increases. Both systems fail when considered with the concept of a minimum wage, for, if a worker fails to achieve agreed targets then effectively by receiving the minimum wage he is paid a higher rate per unit of work (*Fig 2.5*).

Advantages:	*Disadvantages*
1 Direct incentive to increase output.	1 Supervision of quality required.
2 Simple to understand and compute.	2 Targets if given in hours require to be changed following changes in wage rate.
3 Constant wage cost per item (without minimum wage).	

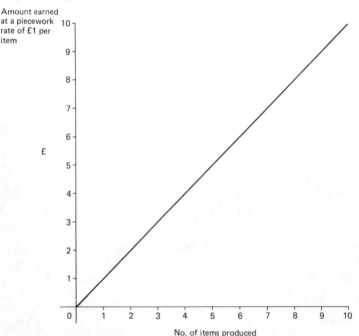

Fig 2.4 Payment system based on piecework

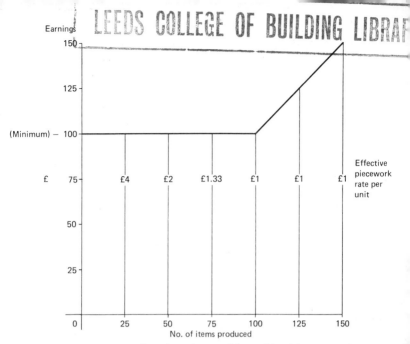

Fig 2.5 Payment system based on piecework but with minimum earnings

2.2.3 HOURS SAVED SYSTEMS

Such systems provide a target in hours for a specified task and on completion the worker is paid the difference between the target set and the hours taken to do the task. If a task carries a bonus target of 20 hours and the basic wage of the worker is £3 per

Table 2.1 Earnings and cost of Hours Saved System

a	b	c	d	e	f
Target hrs	Actual time taken hrs	Hours saved hrs	Daywork earnings £	Bonus earnings £	Cost £
20	16	4	48	12	60
	18	2	54	6	60
	20	0	60	0	60
	22	0	66	0	66
	24	0	72	0	72
		$c = a - b$ min 0	$d =$ $b \times £3/hr$	$e =$ $c \times £3/hr$	$f = d + e$

27

hour, *Table 2.1* compares the earnings of the worker and costs to the company of differing completion times.

This demonstrates the advantages to the worker through reward for the fast without penalty for the slow. Negotiations therefore over the correct level of the initial target are complex.

Advantages:
1 Incentive related to effort
2 Guaranteed minimum wage.
3 Some effective cost control.
4 Quality control more easily supervised.

Disadvantages:
1 More expensive to operate.
2 Favours the faster worker.
3 Requires sound data.
4 Causes problems during the learning process.

2.2.4 GEARED HOURS SAVED SYSTEMS

For many reasons it may be desirable to reward individuals who work more slowly than the average. A geared system initially sets a more attainable target but only pays

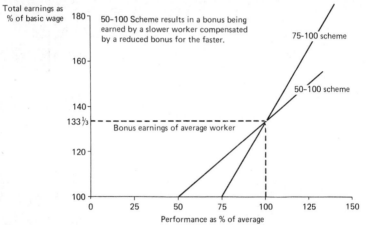

Fig 2.6 Affect on earnings of differing geared schemes

a proportion of the savings back to the worker. *Fig 2.6* illustrates the effect that differing systems have on workers of differing speeds.

Advantages:
1 Useful to accommodate changes.
2 Provides incentives for inexperienced workers.

Disadvantages:
1 Faster workers not fully rewarded.
2 Encourages imposed production ceilings.

2.2.5 GROUP SCHEMES

Many operations cannot be carried out on an individual basis being completed by a number of workers contributing differing skills for differing periods of time. Earnings should therefore be adjusted to accommodate these differences. This is particularly true in the construction industry where craft and other operatives invariably operate in trade gangs. *Table 2.2* illustrates the calculation involved in such a scheme.

Table 2.2 Calculations of bonus within a group scheme

Worker	Bonus Share	Hours Worked	Hours Worked × Share
Chargehand	1½	8	12
Craft operative 1	1¼	16	20
Craft operative 2	1¼	16	20
Operative 1	1	16	16
Operative 2	1	7	7
		63	75

Hours saved = Target 83 hours − actual hours worked 63 = 20 hrs.
Bonus earned = 20 hrs × £3/hr = £60

Bonus per share = $\frac{£60}{75}$ = £0.80 per share

Chargehand	£0.80 × 12 = £ 9.60
Craft operative 1	£0.80 × 20 = £16.00
Craft operative 2	£0.80 × 20 = £16.00
Operative 1	£0.80 × 16 = £12.80
Operative 2	£0.80 × 7 = £ 5.60
Bonus Earned	£60.00

Not all workers output can be measured in normal production units but output might well depend upon their effective work. It is common for such workers to be paid within a group scheme or an average of other workers. Certain plant operations including concrete mixing and distribution may come into this category.

Advantages:
1 Assists in the identification of slow workers.
2 Flexible to cope with differing circumstances.
3 Encourages cooperation.

Disadvantages:
1 Increased computation requirement.
2 Faster workers subsidise the slow.

2.2.6 PLUS RATES

Where additional payments are required irrespective of output, the expedient of plus rates may be used. These are simply payments over and above standard daywork rates. If the additional payment is made weekly or for a particular operation it is often called a 'spot' bonus.

2.2.7 JOB AND FINISH

In this particular system a fixed quantity of work is specified attracting a fixed payment. When the worker has completed the operation he is allowed to leave his place of work without financial penalty.

2.2.8 PROFIT SHARING

A lump sum payment is made to individuals based upon company performance either annually or biannually. The major disadvantages of this scheme are the length of time between the activity and the reward and the assessment of the individuals contribution towards that profit.

2.2.9 PRINCIPLES OF A GOOD REWARD SYSTEM

1   Earnings should be porportional to effort.
2   There should be no restriction to earnings.
3   Targets should be attainable.
4   Targets should remain unaltered unless conditions change.
5   Schemes should be simple to compute.
6   Hold-ups should be prevented by good planning.
7   Standards of quality should be maintained.
8   Time lost through managerial responsibility should be excluded.
9   Targets should be issued for all work before it commences.
10 Gangs should be as small as possible but include all who affect production levels.
11 Payment should be made at regular intervals and as quickly as possible after the task is completed.

2.3 INDIRECT FINANCIAL SYSTEMS

Not all reward systems depend upon the direct payment of money. The use of fringe benefits has been considered the perogative of salaried staff and more specifically higher management. However the need to reward all levels of the workforce in more varied and attractive ways has led to the establishment of a variety of schemes available to all, but particularly for those whose jobs are difficult to measure in the short term or in terms of units of production.

2.3.1 SECURITY OF EARNINGS

To be assured of a basic financial income during the times when the individual is not working is an incentive provided by the state through unemployment benefit, social security payments and state pensions. However by offering a variety of schemes, either contributory or non-contributory, companies are able to offer benefits in excess of the state levels which may enable a worker to maintain a standard of living closer to that which he has established whilst in work. Such schemes include:

1 Sickness benefit.
2 Holidays with pay.
3 Life and accident insurance.
4   Retirement pensions.
5   Private medical insurances.

2.3.2 SUBSIDISED SERVICES

Many facilities may be offered by a company providing necessities or luxuries for which the worker would otherwise have to pay. The rate of subsidy will vary greatly from 100% down. Such schemes include:

1	Refectory services.	5	Transport including buses, company cars, mileage payments and/or fuel.
2	Luncheon vouchers.		
3	Sports facilities.	6	Expense accounts.
4	Social facilities.	7	The use of credit and discount cards.

2.3.3 DOMESTIC EXPENSES

Schemes offering payment of expenses concerning the domestic situation rather than that of work are popular particularly when the circumstances of the job provide little differentiation, a representative working from home for example. Schemes again vary from payment in full to a variety of levels of subsidy and include:

1	Reduced mortgage interest rate.	4	School fees.
2	Subsidised rentals.	5	Telephone accounts.
3	Holidays.	6	Goods voucher systems (operated by most major retail outlets).

2.3.4 OTHER SCHEMES

Two additional schemes have recently become popular although some companies have been operating the first for a number of years. This involves the issue of shares in the company to workers in an endeavour to widen ownership and increase the sense of involvement within the operation of the company.

The second scheme which has been the subject of much official discussion involves the payment of income tax directly to the Inland Revenue by the company so that when a worker calculates his earnings he does not need to reduce the amount by approximately one third to accommodate that lost to taxation.

2.4. NON FINANCIAL SYSTEMS

This chapter has so far provided details of reward systems which, directly or indirectly, provide the means for the worker to increase his standard of living and convenience away from his place of work generally. Such rewards are called 'extrinsic' motivators. The research summarised in 2.1 however clearly indicates that the willingness to work effectively is based upon a much broader set of needs than can be met by such rewards.

The social and physical conditions within the work situation and the job requirements provide the possibility of rewards internal to the individual and are therefore called 'intrinsic' motivators.

2.4.1 THE PHYSICAL WORKING ENVIRONMENT

Human work takes place in a variety of conditions from coal mines and oil rigs to factories and hospitals thus spanning a whole range of conditions. Since the industrial revolution enlightened management has always endeavoured to provide an environment that is as safe and attractive as possible under the circumstances. It has however been necessary for the law to be used to enforce minimum standards to control that management which seeks to exploit the health and safety of its workers in the search for profit. The main areas of interest are:

1	Machinery and plant.	4	Noxious raw materials and by-products.
2	Decoration and repair.		
3	Noise and heat levels.	5	Atmospheric conditions.
		6	Hygiene.

2.4.2 THE SOCIAL WORKING ENVIRONMENT

Most jobs involve interpersonal contact be it purely social with little requirement for cooperation within the work process or essential to effective production. The relationship between superior and subordinate is also important both to the task and to the general environment. A supervisor's attitudes towards his responsibilities and consequent approach can produce an oppressive or cooperative atmosphere.

2.4.3 THE WORK ITSELF

The ability of a job to provide intrinsic motivation has been the subject of scrutiny during the recent past. In the search for efficiency scientific management had:
1 Broken down jobs into small elements to reduce required skill training.
2 Sought to reduce reliance on individuals.
3 Increased effective working hours through the reduction of fatigue by mechanisation.
4 Increased quality control through job standardisation.
5 Increased control through mechanical pacing of operations.

The long term results of this strategy have been:
1 An increase in spoilage, absenteeism and labour turnover through boredom.
2 Demands for high wages in compensation.
3 A reduction in the workers self-image.
4 An unrecognisable contribution by the individual to the end product.
5 Increasing job and social isolation.

The redesign of jobs to produce greater satisfaction has provided the following:
1 The setting of targets in consultation with the worker.
2 The control of preparatory and ancilliary operations.
3 The interlocking of tasks and job rotation in high stress areas.
4 The involvement of the worker in job definitions.
5 The increase of performance feedback.
6 The wider participation within the organisation.
7 The redesign of the approach to work. (The use of work groups instead of assembly lines for example.)

2.5 FINANCIAL REWARD SYSTEMS IN THE CONSTRUCTION INDUSTRY

Operatives working within the construction industry are engaged under a variety of agreements details of one of which can be found in chapter 6, but generally direct financial systems contain the common elements listed below.

2.5.1 DIRECT FINANCIAL SYSTEMS

Weekly earnings If the operative is available for work during the week he will receive:
1 The current weekly standard basic rate at craft or operative level.
2 A guaranteed minimum bonus where this element has fallen below the minimum level.
3 Apprentices, trainees and young operatives are paid a rising percentage of the respective adult rate.

32

Additional payment for continuous skill and responsibility
1 Trade chargehands and gangers.
2 Qualified barbenders and fixers.
3 Scaffolders: trainee, basic and advanced.
4 Timbermen.
5 Mechanical plant operatives.
6 Full-time maintenance fitters.
Occasional payments
1 Work at heights.
2 Furnace and firebrick work.
3 Dirt, inconvenience and discomfort.
4 Simple scaffolding.
5 Mechanical cleaning of stonework.
Extra time working
1 Overtime.
2 Working on statutory holidays.
3 Shiftwork.
4 Nightwork.
Travelling
1 Daily travelling time.
2 Travelling time, not daily.
3 Fare allowance.
4 Lodging allowance.
Tool allowance.
Payments under incentive and/or productivity agreements.

2.5.2 INDIRECT FINANCIAL SYSTEMS

Holidays with pay Provided the operative is available for a specified minimum number
of working days within a pay week, the employer affixes a holiday credit stamp which
can be exchanged at the appropriate time for payment in respect of annual and public
holidays.
Sickness and injury Adult operatives absent from work for more than three
consecutive days as the result of sickness or injury receive a fixed sum payment per day
on the production of a medical certificate.
Death benefit cover If an operative, with the required attendance between the ages of
16—65 dies whilst in employment or at any age if the death is the result of an accident
at work or travelling to and from work, a lump sum payment is made to the spouse or
dependents.

2.5.3 ESTABLISHING A DIRECT FINANCIAL INCENTIVE SCHEME

Detailed consultations will be required before a direct financial incentive system can be
introduced. Consultations should take place not only within the organisation between
those divisions which will become involved in its operation; site management,
estimating, costing and planning, but also consultations external to the company with
representatives of the trade union or trade unions which represent the majority of the
men. When the objectives have been clarified and the responsibility for the development
allocated the following stages may be followed:

1 Select system.	6 Prepare the short term plan.
2 Establish the operating rules.	7 Issue the weekly plan and the targets.
3 Identify the tasks to be targeted.	8 Allocate the labour time to the tasks.
4 Calculate the quantities of work involved.	9 Maintain the bonus records.
5 Set the targets.	10 Issue feedback information.

When details of the scheme are in preparation, consultations with the trade unions can be brought down to a local and site level. A clear effort must be made to sell the scheme and this involves the establishment of a disputes procedure which may be accommodated within existing arrangements.

The installation of a direct financial incentive scheme affects many people both inside and outside the contractor's head office. If it is to be established with the goodwill of those people and thus stand a chance of meeting its objectives it must only be designed after negotiation and careful consideration of the people involved.

2.6 NON-FINANCIAL REWARD SYSTEMS IN THE CONSTRUCTION INDUSTRY

2.6.1 THE PHYSICAL WORKING ENVIRONMENT

The construction industry offers a variety of working conditions, but, because the majority involve commencement of work below ground level and then on an incomplete structure, controlled conditions are difficult to achieve. Work is often required in conditions of cold, wet and wind but also offers the advantages of outdoor work on a warm sunny day.

Whilst good planning can alleviate some of the worst excesses of working conditions it can never remove entirely the exposure factor. It must therefore be accepted that alternative reward systems should provide facilities which will recompense the operative and at least provide services which will facilitate full recovery from the effects of exposure. Such facilities should include:

1 Protective clothing.	3 Sanitary facilities.
2 Washing facilities.	4 Changing and clothes storage facilities.

Whilst these aspects are regulated partly by law and partly by agreement they have historically been neglected areas and there must be considered to be room for improvement particularly having regard to recent advances in mobile accomodation units.

2.6.2 JOB SECURITY

It is generally accepted that the building industry is a casual based industry offering little in the way of continuity of employment. Many reports, however, suggest such discontinuity is self-induced and if measured on an industry basis and not a company basis then the record is considerably improved.

The reasons for such discontinuity are varied. Firstly, there is a large amount of voluntary movement particularly amongst the active single members of the industry who recognise that their earning capacity will diminish as they grow older, dependent as it is on productivity bonuses. Secondly, operatives tend to work within a geographic area which may not correspond with a particular company's area of operations. It is also argued that the largest clients of the industry are funded by public money and are therefore a prime target for governments wishing to regulate the money supply.

Whilst this casual nature of employment provides a high degree of flexibility required to cope with an uncertain workload (compounded by competitive tendering as a means of allocating work), it is expensive in terms of efficiency and lowers the public image. It also reduces the incentive for companies to attempt long term training programmes and benefits such as pension schemes.

For these reasons a central employment agency for building workers along the lines of the National Dock Labour Board has been suggested as a move towards decasualisation. However as the scheme attained political significance it has of late become dormant.

2.6.3 HEALTH AND SAFETY

After mining and fishing, construction ranks third in the accident league table, with a death occuring every working day. The Health and Safety at Work Etc. Act 1974 became the principal legislative force providing for the introduction of Statutory Instruments eventually to supercede all existing legislation. The main purpose of the Act is to widen the responsibility for health and safety to include all those who work in the construction process. This it hopes to achieve by establishing mechanisms of participation by both management and production workers.

The relationships between safe methods of working and productivity are in the short term complex, but the total cost of accidents on site leaves no doubt as to their disruptive nature particularly when a death is involved. It is often suggested that workers are oblivious to their own safety and, in certain circumstances, through contemptuous familiarity or a desire to increase earnings, they will deliberately place themselves in jeopardy. However it must be clear to all that productivity must never be achieved through knowingly adopting unsafe practices, and that bonus targets must be based on safe methods of working which must be rigorously controlled.

In addition to killing men through accidents, the construction industry also kills men slowly. Inadequate eating facilities, insanitary conditions and a general lack of knowledge regarding suitable clothing all result in an increased possibility of a partial disablement towards the end of a construction workers working life. Respiratory and muscular difficulties have been shown to be more common amongst older construction workers than the average industrial worker of similar age. These conditions are achieved within the law because, although the use of safety helmets is often mandatory by voluntary agreement, the wearing of specially designed clothing is still a rare occurrence. The chill factor of a cold wind blowing across damp denim, reduces body temperature to hypothermic levels reducing productivity and eventually incapacitating the operative.

It appears clear that these factors affect little the motivation of the young fit entrant to the industry, but how does such knowledge affect recruitment and how many skilled men are lost to the industry because of a growing awareness of such dangers?

2.6.4 THE SOCIAL WORKING ENVIRONMENT

Construction is by nature a group activity with the majority of operations being carried out by gangs of operatives who are in the main self-selected and reasonably permanent. This creates established leadership patterns and general satisfaction with the immediate social setting of work.

The next level of supervision is that of trade foreman who, being almost invariably promoted from that trade, experiences few communication problems and commences problem solving from an identical background. Supervision is generally not oppressive because of the physical layout of the site and this allows the craft operative certain freedom which is appreciated in the main.

The lack of formally acquired supervisory skills by any other method than experience is now recognised and schemes are available as detailed later to provide established supervisors and managers with the theoretical background with which to measure their on site effectiveness.

2.6.5 THE WORK ITSELF

The image of construction work, particularly in the eyes of those concerned with career advice, is that of heavy, dirty manual labour lacking in glamour and skill with limited career prospects. These views along with the factors previously mentioned reduce the status of the construction worker.

Paradoxically however, detailed investigation of the work situation will show that it is capable of providing satisfaction where mass production fails. The building worker is part of a team whose constitution and objectives are ever changing. Each project is a permanent construction required to meet genuine basic needs of society and generally will remain for many decades as a permanent monument to the skill of its constructors.

However the traditional skill based organisation is threatened from a variety of sources:
1 The simplification of building design for economy.
2 The increase in labour costs.
3 The increasing sophistication of machinery.
4 The economics of prefabrication.
5 The development of new materials.

Mistakes have been made in the past in the search for fast, cheap building systems, but short term low cost schemes have proved expensive in the light of experience, both technically and socially. Society requires buildings at a cost which it can afford but it must be aware of the cost not only in financial terms but also in human terms.

BIBLIOGRAPHY

The Chartered Institute of Building. Annotated bibliographies
10 Incentives
64 Incentive schemes for maintenance
CIOB, Ascot
The Chartered Institute of Building. *Occasional Paper 10; Financial incentives* — Do they work?, CIOB, Ascot
The Chartered Institute of Building. Occasional Paper 19 Worker motivation in building, CIOB, Ascot
Geary, R., Work study applied to building, Godwin, (1962)
Harris, F. and McCaffer, R., *Modern construction management,* Crosby Lockwood Staples, London (1977)
National Joint Council for the Building Industry. *Working Rule Agreement* (1982 edition) NJCBI London

EXERCISES

RECALL QUESTIONS

1 List the headings under which a construction worker's direct earnings may be analysed

(a) ....   (e) ....
(b) ....   (f) ....
(c) ....   (g) ....
(d) ....

2 Identify the category of motivator by using the letter E for extrinsic and the letter I for intrinsic
   (a) Daywork earnings   ...   (e) Working conditions   ...
   (b) Feedback of results ...   (f) Pension scheme   ...
   (c) Nature of supervision ...   (g) Job satisfaction   ...
   (d) Sick pay   ...   (h) Profit sharing   ...

3 Name the direct financial reward system described below
   (a) Payment of hourly rate based on attendance ....
   (b) Payment on the basis of units of work completed ....
   (c) Payment of the whole difference between time taken and target ....
   (d) Extra to the standard hourly rate paid to all   ....

4 Complete the following table by identifying the need hierarchy as proposed by Maslow and by giving one example of each level

Need	Example
S	....
E	....
B	....
S	....
P	....

5 By completing the boxes in the diagram below, identify the factors involved in the process of motivation.

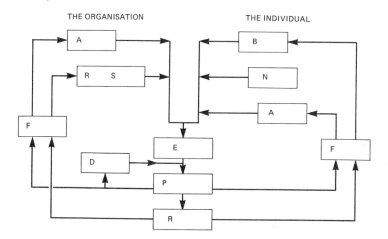

THE ORGANISATION   THE INDIVIDUAL

QUESTIONS REQUIRING SHORT ESSAY ANSWERS (15–20 minutes)

1 How did the findings of the Hawthorne Investigations call into question the views of scientific management?

2 List the principles of a good direct incentive scheme.

3 Identify the categories of direct financial reward systems giving examples of their effective use in the construction industry.

4 A substantial part of a construction worker's earnings are often the result of a financial incentive scheme. Outline the advantages and disadvantages of such a situation to both parties.

5 List the relative merits of the casual basis of employment in the construction industry.

QUESTIONS REQUIRING LONGER ESSAY ANSWERS (30–45 minutes)

1 Describe the procedure for the establishment of a financial incentive scheme within a construction company.

2 How could the findings of either:
(a) Herzberg's Motivation Hygiene Theory, or
(b) McGregor's Theory X and Theory Y
be used to increase job satisfaction in the construction trades?

3 Explain the part played by non-financial reward systems in the process of motivation.

4 Discuss the proposition that the building site is a dirty, dangerous and demoralising place of work.

5 Using the classification Rational, Economic, Social and Self Actualising Man explain how the attitudes of the organisation affect the implementation of reward systems.

3 Personnel management

3.1 Operational areas	3.1.1 Organisational analysis 3.1.2 Employee remuneration 3.1.3 Administration and records 3.1.4 Industrial relations 3.1.5 Employee services 3.1.6 Manpower development and training 3.1.7 Manpower planning 3.1.8 Employment

3.2 Manpower planning	3.2.1 Establish company plans 3.2.2 Assess existing manpower 3.2.3 Assess external factors 3.2.4 Programme staff development 3.2.5 Produce and implement the manpower plan 3.2.6 Evaluate forecasts

3.3 Recruitment	3.3.1 Job analysis 3.3.2 Job descriptions 3.3.3 Personnel specification 3.3.4 Attracting suitable candidates

3.4 Selection	3.4.1 Appraisal 3.4.2 The selection interview 3.4.3 Objective testing

3.5 Placement	3.5.1 Decision 3.5.2 Induction 3.5.3 Follow-up

3.1 OPERATIONAL AREAS

Personnel management is that function concerned with people rather than finished products. It is related to individuals and groups; to inter-relationships and to the ways in which an individual can contribute to the development of himself and the organisation. The aim of personnel management is to provide an environment within which all employees may realise their true potential. It may be going too far to say what is good for the individual is good for the firm but there is no doubt that organisational effectiveness can be increased by the adoption of a suitable developmental philosophy.

As with other functions, personnel management can be carried out by specialists or be regarded as an integral part of general management. Which is adopted by a particular company depends upon the size and company policy, but it must be recognised that the employment of a specialist no more guarantees effectiveness than does the employment of a salesman guarantee sales.

The following represent what may be considered to be the operational areas of personnel management and are functions which must be carried out by any organisation which wishes to prosper in a changing environment.

Organisational analysis The assessment of the appropriateness of the organisational structure and working methods, and the way they assist in the meeting of company objectives. (Chapter 1)

Employee remuneration The development and maintenance of appropriate reward structures. (Chapter 2)

Administration and records The maintenance of adequate and proper records to ensure adequate knowledge of the current situation and to provide statistical data for the future. (Chapter 3)

Industrial relations The establishment of proper and adequate relationships between management and production employees on both individual and group bases. (Chapters 4, 5, 6, 7)

Employee services The establishment and maintenance of appropriate standards in relation to health, safety and welfare of all employees. (Chapter 8)

Manpower development and training The assessment of training needs related to both the company and the individual. (Chapter 9)

It is indicated that these topics are covered elsewhere in the book and this section deals specifically with that area perhaps most central to organisational effectiveness, that is the manning of the structure with appropriate personnel. Success in this area depends upon effectiveness in the following functions.

Manpower planning The assessment of the appropriateness of present manpower to meet present and future needs.

Employment The taking of measures to ensure a balance of manpower between the supply and the needs of the organisation. This can be further subdivided into the functions of: Recruitment; Selection; Placement.

3.2 MANPOWER PLANNING

The aims of manpower planning are:
1 To obtain and retain manpower correct in both quantity and quality.
2 To obtain maximum utilisation of existing manpower resources.
3 To anticipate potential surpluses and deficits and identify resultant problems.
Reference has already been made to the casual nature of employment in the
construction industry. Planning in an uncertain environment is most difficult, but
opportunities can only be taken when adequate forecasting has resulted in the
availability of the required resources.

Formal enquiries into conditions within the construction industry have made
accusations of short term exploitation of operatives resulting in:
1 Low productivity.
2 High labour turnover rate, related both to the company and the industry.
3 Reduced motivation to train and plan for the future.
Successive legislation has made such exploitation more difficult and, although labour-
only subcontracting has proved for some to be the solution, many contracting
organisations recognised manpower planning as a necessary and important part of
company planning.

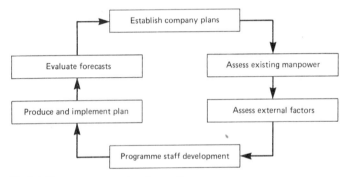

Fig 3.1 The manpower planning cycle

Fig 3.1 indicates the six stages inherent in manpower planning which should not be
considered as isolated, once and for all activities. Each stage must interlink with the
others to form a series of cyclic activities reacting to changes both inside and outside
the company.

3.2.1 ESTABLISH COMPANY PLANS

Manpower planning can only be effective if it is carried out within the context of the
overall plan and objectives of the company. Equally those objectives will not be
reached if they do not take into account the likely future availability of manpower
resources. Manpower planning must therefore be considered as part of company
planning.

Large companies will no doubt possess the necessary expertise in this area but
smaller companies might well consider the need for external assistance. However the
objectives are generated, it is only the principals of the company who can make the

necessary fundamental decisions. The company plan will develop from the answers to a large number of questions which might include:

1 What are the objectives of the company and how are they inter-related? To what extent do they specify profitability, growth, diversification etc?
2 Which is the company market? Is it expanding or contracting? What other opportunities exist? How is the opposition performing?
3 What information exists to assist the decision making process?
4 What are the current strengths and weaknesses of the company?

Following the establishment of a company plan, departmental objectives can be set and a course of action plotted. An analysis of the necessary skill requirement will enable a budget to be formulated and at this stage an invaluable contribution can be made by joint consultation with existing post-holders.

The construction industry is a composite of many markets, the major division being that between building and civil engineering. Distinction can also be made between the technology of construction (traditional – prefabricated); the type of work (new – refurbishment); the service offered (build only – design and build) and the financial value of the contract. Each different market determines the manpower requirements for managerial, technical and administrative staff.

Because the similarities are greater than the differences and because of the uncertainty of competitive tendering many firms find themselves operating in more than one market or moving from one to another as the availability and nature of work change. If the contractor is not able to recognise the differences in skill requirement and has not forecast his commitment to that market, then, just as he might operate sophisticated plant incorrectly, he will not be utilising his manpower resources to their greatest effect.

3.2.2 ASSESS EXISTING MANPOWER

Given a company plan and knowledge of the manpower needs for its implementation, a thorough analysis of existing manpower should take place. This involves the gathering of information from a variety of sources, for example, from personnel files, application forms, training records and appraisal exercises.

All planning depends upon the accuracy of basic information and a good record system, in terms of individual detail and general statistical data, will increase the validity of labour movement forecasts.

Personnel records should be maintained for all employees. This may be done on a card index system in the smaller firm or on a computer system with full data retrieval in the larger firm. *Fig 3.2* gives an example of a record card.

Statistical information can now be collated to give analyses of manpower as required and may relate to:

Age distribution;	Job experience;
Length of service;	Salary/wage patterns;
Trade or position;	Performance;
Education level;	Reasons for leaving.
Training;	

As all employees will be the subject of analysis it will be of greater use to group them into occupational categories. Production workers can be analysed by trade; supervision and managerial by level and specialists by expertise. Thus joiners will be assessed separately from bricklayers; site managers from contracts managers and quantity surveyors from estimators.

The information once assembled can be examined to discover any immediately

```
┌─────────────────────────────────────────────────────────────────────────┐
│  EMPLOYEE            NAME (Mr.Mrs.Miss).........................          │
│  RECORD             ADDRESS.....................................          │
│  CARD                .........................TEL NO. .........           │
├───────────────────────────────────────────────────────────────────────── │
│ INITIAL APPOINTMENT ..................................................... │
├────────────┬──────────┬─────────────┬────────────┬──────────┬─────────────┤
│ Marital    │ Date of  │ Nationality │ Date of    │ Salary   │ Registered  │
│    Status  │   Birth  │             │ Employment │          │ Disabled    │
│            │          │             │            │          │ YES/NO      │
│            │          │             │            │          │             │
│            │          │             │            │          │ NO. ....... │
├────────────┴──────────┴─────────────┼────────────┴──────────┴─────────────┤
│ EDUCATION                           │ TRAINING                            │
│                                     │                                     │
│                                     │                                     │
│                                     │                                     │
├─────────────────────────────────────┴─────────────────────────────────────┤
│ PREVIOUS EMPLOYMENT                                                         │
│                                                                             │
│                                                                             │
└─────────────────────────────────────────────────────────────────────────┘
```

FACE

```
┌─────────────────────────────────────────────────────────────────────────┐
│ STATUS AND SALARY REVISION                                                │
├──────┬────────┬───────────────────────────────────────────────────────────┤
│ Date │ Salary │ Reason for variation                                      │
│      │        │                                                           │
│      │        │                                                           │
│      │        │                                                           │
│      │        │                                                           │
│      │        │                                                           │
│      │        │                                                           │
├──────┴────────┴───────────────────────────────────────────────────────────┤
│ FURTHER EDUCATION AND TRAINING                                            │
├──────┬───────────────────┬─────────────────────────────┬──────────────────┤
│ Date │ Place of training │ Nature of training          │ Result           │
│      │                   │                             │                  │
│      │                   │                             │                  │
│      │                   │                             │                  │
│      │                   │                             │                  │
│      │                   │                             │                  │
└──────┴───────────────────┴─────────────────────────────┴──────────────────┘
```

REVERSE

Fig 3.2 A typical small company record card

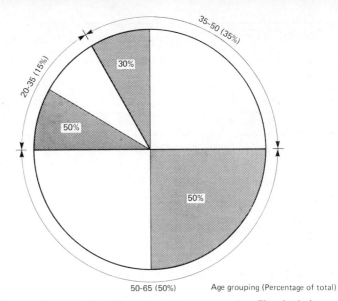

Fig 3.3 Pie chart representing site managers by age grouping. The shaded portion represents that proportion of each age grouping which could normally be expected to leave within the period under consideration

obvious problem in either the short or long term. *Fig 3.3* shows a suggested age distribution of site managers within a company indicating those who are likely to leave within a given period. The implications of this information can be considered in relation to required replacements in the future and also the preferred age range of recruits.

This exercise has so far indicated the problem areas and the following questions can be applied to reach solutions to those problems:
1 Is there potential for the improvement of existing manpower?
2 Which individuals have such potential?
3 Can a formal training programme improve the situation?
or
4 Do we need external recruitment?

3.2.3 ASSESS EXTERNAL FACTORS

Although a company with adequate manpower planning and development policies might find it rarely has to buy in skills or expertise, the consideration of the effects of external factors within the job market cannot be ignored. Consideration must be given to local and national events which might well effect recruitment policy in the future. This is particularly important in respect of those skills which have been identified as being crucial to the meeting of company plans. Information for this assessment will have to be sought from outside agencies both local and national.

The response of the firm to changes in the job market will depend upon the answers to the following questions:

44

1 How are local population densities changing?
2 What are the local levels of employment?
3 Is there a national population shift?
4 Which industries compete with construction for labour?
5 How are they likely to perform in the future?
6 Which are our most profitable sources of recruits in each category?
7 How are skill requirements changing?
8 What effect is legislation likely to have on employment?
and
9 How is the economy likely to perform?
Such questions are, of course, doubly important to the construction industry. They are not only relevant to manpower planning but also provide marketing information used in the establishment of company plans.

3.2.4 PROGRAMME STAFF DEVELOPMENT

The problems isolated by the previous stages of manpower planning can be differing combinations of the following:

1 Labour shortages or surpluses.	5 Imbalance in age groupings.
2 High labour turnover in certain jobs.	6 Unacceptable levels of absenteesim.
3 Poor performance in certain jobs.	7 Redundancy of skills and experience.
4 Unacceptably high rate of accidents.	

The solution to some of these problems may only be reached through changes in policy and recruitment, but a programme of training carried out after careful counselling might overcome others and allow solutions to be sought within existing resources.

Training may be given at three levels each of which is designed to meet separate objectives.
1 *Induction training* which familiarises employees with their jobs and with the company allowing them to settle much more quickly and avoid misunderstandings and frustrations.
2 *Skill training* which is designed to improve performance of current tasks.
3 *Development training* which provides the mechanism for personal development within a career structure.
The traditional assumption is that skill training relates to production workers and development training to managerial and technical workers. It has however been shown that increased motivation can result from the acceptance of responsibilities and the aquisition of new skills and knowledge. The appraisal of potential within individuals is in itself a skill which might well form the basis of a training course for management and supervisory workers.

3.2.5 PRODUCE AND IMPLEMENT THE MANPOWER PLAN

A record of future manpower availability can now be compared with the projected need in the manner described:

1 Decide on labour grouping to be forecast.	4 Identify the manpower gap.
2 Identify additional manpower requirements.	5 Repeat for other groupings.
3 Estimate manpower losses.	6 Combine data and formulate the plan.

Fig 3.4 shows how such a computation can be made for a particular group. It is

OUTLINE MANPOWER PLAN FOR SITE MANAGERS   1982–1984

1.	Site Manager requirement	December 1984		65
2.	Site Manager employment	January 1982	55	
	Projected turnover			
	(a) Retirement	3		
	(b) Promotion/transfer	2		
	(c) Dismissals	1		
	(d) Withdrawals	5	11	44
3.	Manpower Gap			21

Fig 3.4 Specimen outline manpower plan

unlikely that there will be an exact match between availability and requirement so one of the three following situations will prevail:

1 *Sufficient manpower is employed* The company will only need to monitor the situation to ensure that changing circumstances do not alter the position.
2 *Excess manpower is employed* Consultations must commence with employee representatives regarding the reduction of manning levels. This may be done through natural wastage, retraining, reassessment of objectives or, as a last resort, redundancy.
3 *Additional manpower is required* An operating plan for recruitment, promotion, retraining and induction training must be developed.

3.2.6 EVALUATE FORECASTS

Planning provides the opportunity for control, but if constant monitoring of the actual situation is not undertaken, and comparison with the prediction made, then the plan will no longer be valid.

All forecasts made during the manpower planning process should be verified and corrections made. These will include:

1. Predicted changes in activity.
2. Trends.
3. External factors.
4. Manpower losses.

Any amendment to these forecasts must be carefully considered and its effect on the total manpower plan calculated.

3.3 RECRUITMENT

It has been shown in the previous section on manpower planning that recruitment may be required by an organisation for many reasons:

1 The existing postholder may have withdrawn, retired, been dismissed, promoted or transferred.
2 Expansion may have created an additional post of an existing defined job.

3 A new job may have been created by diversification.

4 A new job may have been created by a new technology or through reorganisation.

The recruitment phase can be considered to be the securing of a supply of qualified candidates for a post within an organisation and comprises the following stages:

1 Job analysis
2 Job description
3 Personnel specification
4 Attraction of suitable candidates

3.3.1 JOB ANALYSIS

Job analysis is the scientific study and statement of all the factors concerning a job, which reveal its total content. It can best be carried out by the systematic application of the following questions:

What are the outline details?
The objectives.
The major duties and responsibilities.
The major parameters — age/salary/experience.
The major difficulties.
The career prospects.

Are the objectives still valid?
Jobs are changing concepts, a vacancy presents an opportunity to re-think its basic purpose.

Do we really need to recruit from outside our organisation?
External recruitment is costly, difficult and frought with dangers. Is it really necessary?

Is the job a saleable proposition?
Does the job provide sufficient intrinsic and extrinsic rewards to motivate?

Can the job be changed to attract better people?
If the answer to the last question is no then the job will have to be redesigned.

3.3.2 JOB DESCRIPTIONS

The outline details of the job can now be expanded by observation and by consultation with existing postholders, superiors and subordinates and available records. The following headings present a checklist to ensure that every important fact related to the job has been recorded:

1 Job title and grade.
2 Location.
3 Duties.
4 Responsibilities.
5 Working conditions — physical
    — social
6 Training.
7 Remuneration.

3.3.3 PERSONNEL SPECIFICATION

Having recorded the nature of the job in detail it is now necessary to obtain a picture of the ideal candidate. One of the most widely used classifications of human attributes is the Seven Point Plan developed by Professor Alec Rodger in the early 50's.

1 *Physical make up* Defects of health and physique of occupational importance, appearance, bearing and speech etc.

	Essential	Preferable	Contra-indication
Physical make-up			
Attainments			
General intelligence			
Special aptitudes			
Interests			
Disposition			
Other essential details			

PERSONNEL SPECIFICATION

Fig 3.5 Typical personnel specification sheet

2 *Attainments* Type and extent of education and occupational training, achievements.
3 *General intelligence* Possession and display of general intelligence.
4 *Special aptitudes* Aptitudes for words, figures, mechanics etc.
5 *Interests* Intellectual, practical, constructional, social, artistic etc.
6 *Disposition* Acceptability to others, steadiness, dependance, reliability etc.
7 *Circumstances* Present and past domestic and personal circumstances.

The presentation of the factors does not imply relative importance, and the information is not always of the type which can be gained through direct questioning. It does however allow a picture of the ideal candidate to be developed against which applicants can be measured. *Fig 3.5* illustrates a form which should be completed with objective detail related to the job.

Each of the seven points with the exception of *circumstances* is analysed under the following headings:

1 *Essential* Positive attributes which are absolutely necessary for the successful candidate in order that the job may be carried out effectively.

2 *Preferable* Positive attributes which are not essential but which would lead to superior performance.
3 *Contra indications* Positive attributes which automatically disqualify a person from the job.

Domestic circumstances are only included in such a specification when they have occupational importance, otherwise they are only required to explain or put into perspective the other six elements of the plan.

The detail thus built up provides a concentrated list of those required characteristics and those which must be avoided and will form the basis of an advertisement designed to attract suitable candidates.

3.3.4 ATTRACTING SUITABLE CANDIDATES

People who are changing their jobs tend to follow similar patterns and successful recruitment requires that a company takes note of this behaviour.

Occupation	*Pattern*
Operative	Enquiries on site, word of mouth, Department of Employment Job Centre, Contact also through records of previous employees.
Craft operative	All the above plus local advertisements and trade unions.
Young trainees	Youth employment service, schools, relatives of existing employees, direct approach to company.
Technical/ professional staff	Department of Employment Professional and Executive Recruitment, centres of further and higher education, national, local and trade/professional press.
Senior management/ top people	Personal introductions, consultants, national and professional press.

The form and position of any advertisement must be matched to the job. The information must be adequate but not excessive, it must be clear and command attention. It must also specify the manner of response i.e. Ring Mr. Jones on freephone . . . . , send for an application form and job description, call on site and ask for Mr. Smith etc.

The following is a list of positions where advertisements may be exhibited to good effect:
1 Company notice boards and journals.
2 Boards at site entrance.
3 Lineage advertisements in local/area/national newspapers trade magazines and professional journals.
4 Block or display advertisements in the same.
5 Department of Employment Job Centre and Professional and Executive Recruitment services.
6 Local radio.
7 Local freesheet.
8 Institutions of further and higher education.

An advertisement particularly for external display should contain the following:
1 Introduction — meaningful headline with compulsion to read on.
2 The company — provide a brief understanding and promote job against competition.
3 The job — brief details of essential elements.

4 The requirements — summary of essential characteristics.
5 The incentives — details of major aspects of the reward system.
6 The action to be taken.

Some methods of advertising are free and some are very expensive, so the objective must be to obtain the maximum qualified response for the minimum outlay. The strategy for success will depend upon the nature of the job being offered but a careful record of past responses will eventually provide management with data which will increase the effectiveness of the chosen strategy.

3.4 SELECTION

If the recruitment programme has been carried out effectively more than one candidate possessing all the minimum required qualifications will apply. Selection has therefore to take place. Selection is a matching process whereby the abilities, aptitudes, interests and personalities of the applicants are matched against the job description and personnel specification. The objective is to select the best person for the job so that there are two areas of failure:
1 By picking the wrong candidate after failing to expose disqualification.
2 By missing the right candidate after failing to expose positive qualities.

The effort and expense involved in the selection process must be related to the cost of such failure in either area. Failure in certain jobs might be more tolerable than in those which require extensive training and involve long term complex decision making. However the costs of employment escalate and current legislation places restrictions on managements which may considerably affect the traditional 'hire and fire' policies of construction site management.

For those areas where failure is not tolerable a formal system of selection must take place. This will involve two stages:
   1 Appraisal.
   2 The selection interview.
and possibly a third
   3 Objective testing.

3.4.1 APPRAISAL

Direct matching of a candidate against a personnel specification, point by point, is the simplest method of selection providing the data is correct and that the result is the clear identification of one superior candidate.

This process can be used for the employment of site operatives when the essential requirements of an appropriate skill level, a recent successful employment record and acceptance of the conditions of employment are readily identifiable. For other jobs it might well be just the first step when more than one satisfactory candidate has applied.

The process requires information and the most efficient presentation is the use of an application form generally designed to meet the needs of a group of similar positions. Young trainees will be required to furnish detailed information about their schooling whereas the form for mature candidates will concentrate on previous job experience. A typical application form is given in *Fig 3.6*. The advantages of the use of a standard form are:
   1 The applicant knows precisely what information is required.
   2 The appraisor can make direct comparisons within the various areas.

The form must not however stifle initiative and the applicant should be encouraged to extend on additional paper if necessary the section related to his motivation. The stages of the matching process are:

1 Take up references and make checks on the declared information.

2 Check essential requirements under each heading.

3 Compare candidates within each heading.

If there are more than a required minimum of candidates at this point who meet all the essential requirements, the desirable qualifications may be used to produce a short list or alternatively brief preliminary interviews may be arranged.

Throughout all stages an established administrative procedure must monitor progress and advise candidates accordingly. Acknowledgement of receipt, notice of rejection and arrangements for interview must be made with tact and thoroughness. This process is the window through which the company is viewed by outsiders and errors and omissions in administration help create a reputation which may reflect on company service.

3.4.2 THE SELECTION INTERVIEW

The interview as a selection aid has many advantages:

1 It is a convenient assessment situation providing face to face communication and instant feedback.

2 Candidates accept and will indeed expect it.

3 It allows the candidate to make an assessment of the company and the position.

There is evidence to suggest that the selection interview is a poor predictor of future performance but frequently the fault lies with the skill of the interviewer rather than the process. The objectives of the interview are:

1 To establish the suitability of the candidate for the job.

2 To present an accurate picture of the job offered.

3 To provide a fair hearing for all applicants.

In order to achieve these objectives three stages can be identified; planning, conduct and follow through, the latter stage is dealt with under section 3.5.

Planning The first aspect of planning, already mentioned, is the administrative process of arranging the date, time and venue of the interview. Sufficient notice must be provided to the candidate and, where possible particular requirements met. For instance a time within a lunch break or shortly after the end of a working day may be preferable to either or both sides.

The room in which the interview is to be held must be prearranged with adequate and suitable furnishings and a waiting area, away from general circulation, provided if there is to be more than one candidate. A cloakroom with the standard facilities should also be made available particularly where candidates have travelled long distances.

Near the day of the interview the application form should be perused and a general plan and timetable prepared. Answers must be found to the following questions:

How can rapport be quickly established?

What questions of fact need to be clarified?

What questions are required to expose the underlying motives?

Which areas need tactful exploration?

Conduct If the preparation has been adequate then the introductory phase of the interview can quickly give way to the more formal area of data collection. The interviewer, by the use of open exploratory questions, should encourage the candidate

<table>
<tr>
<td colspan="2">

APPLICATION FOR EMPLOYMENT

PRIVATE AND CONFIDENTIAL
</td>
<td colspan="2">

POSITION APPLIED FOR

....

FULL TIME/PART TIME
</td>
</tr>
</table>

PERSONAL PARTICULARS		

Surname ....   Nationality ....

First names ....   Place of birth ....

Maiden name ....   Date of birth ....

Address ....   Marital status ....

....   Children   Girls ages .. , .. , .. , .. ,

Tel.No. ....Post Code ....   Boys ages .. , .. , .. , .. ,

Car Owner   YES / NO   National Ins. No. ....

Driving Lic.   YES / NO

          CAR / HGV   Registered Disabled   YES / NO

Endorsements   YES / NO   Registered Disabled Number ....

GENERAL EDUCATION   (Please give details of schools since the age of 11)

Dates		Name of school	Type	Examinations sat	Result
From	To				

FURTHER EDUCATION AND TRAINING

Dates		Name of Institution	Type	Examinations sat	Result
From	To				

Fig 3.6 A typical application form

to talk about his past experiences, his present opinions and his future ambitions, thus revealing those details which can never be elicited by an application form.

The interviewer might like to make notes during the process of the interview and, although this should be done openly, the candidate should not be able to detect what is being written. Care must be taken not to let this break down the relationship established particularly when related to areas of the candidate's life which could be considered embarassing. A set down pen and sympathetic attention will encourage such revelations.

EMPLOYMENT HISTORY

Name and address of past employers	Dates of employment	Job title	Reason for leaving

REFEREES

Name	Address	Position
1		
2		

LETTER OF APPLICATION (Continue on additional sheet if required)

I declare the above information to be correct and understand that deliberate omission may lead to cancellation of agreements.   Date ....
Signed ....

REVERSE

When questioning on both sides is complete the interview should be terminated deliberately but not rudely and should end with a statement which will inform the candidate of the timing of the decision. 'Thank you for coming, we will advise you of our decision by Thursday the first of July.' might be a typical closing remark over a handshake. The objective is to prevent the candidate being left in limbo awaiting a telephone call or the post.

3.4.3 OBJECTIVE TESTING

It has been suggested that the interview is a poor predictor and that testing is one method of achieving a greater success rate. The advocates of testing suggest that:
1 Where available, tests are more objective than interviews.
2 Tests are better for the young who do not have industrial experience.
3 Intelligence tests have value in the appointment of supervisory, managerial and clerical personnel.
4 Aptitude tests can be devised to assist in the selection of machine operators and process workers.

These advantages can only be gained if the correct test is used and the tester is suitably qualified to administer the test and interpret the results. Three factors must be considered before any test results are used:
1 *The relevance* – does the test predict the behaviour required in the job?
2 *The validity* – do people who do well in the test do well in the job and vice-versa?
3 *The reliability* – are the results consistant over a period of time?
Many different types of test are available and each has significance in different types of job selection.

Job sampler tests This is the most direct test in which the candidate is given the actual job to do but in isolated conditions. Skilled operatives and clerical workers can be tested in this manner and architects and designers can be asked to provide portfolios of their work.

Intelligence tests Intelligence testing has a long history with reference to education but is not generally widely used in industry. There is considerable debate about what is actually being tested, and about the validity of any result expressed as a single measure. The tests generally are a combination of numerical, diagrammatic and verbal questions set either with a time limit or in ascending order of difficulty.

Aptitude Tests Aptitudes are the potential to perform the special activities required by certain tasks and these tests are designed to predict whether such potential is present. Many different tests can be administered in a battery and the results issued as an 'aptitude profile' covering the following areas:
1 *Verbal* – the use of words and language.
2 *Numeric* – the use and manipulation of numbers.
3 *Spacial* – the use and perception of shapes.
4 *Mechanical* – The understanding of the transmission of movement.
5 *Manual dexterity* – The control of muscular activity, perception and coordination.
6 *Clerical* – the ability to check and classify words, figures and symbols.

Attainment tests Aptitude tests predict how well an individual might perform a task in the future, attainment tests attempt to quantify or evaluate a current skill level. Some attainment tests are considered aptitude tests for the next stage of skill acquisition. GCE '0' levels for example may be used to select students for an 'A' level course.

Personality tests The application and interpretation of personality tests is a complex skilled activity. The relationship between a job and any personality facet must be very carefully validated before the results of any such tests are used as a predictor of performance.

3.5 PLACEMENT

The final stage in the employment of new personnel by an organisation involves the steps of:

1 Decision.
2 Induction.
3 Follow-up.

3.5.1 DECISION

No matter how effective are the testing and interview stages they can only provide data upon which a decision must be made. It is the quality of that decision which will determine the successful attainment of the original objectives. The involvement of more than one person in this process increases the chances of success as will the following routine:

1 Each candidate should be assessed by each assessor using an alphabetic grading on each key factor of the personnel specification in addition to interpersonnel skills, flexibility, emotional adjustment and motivation.
2 Each assessor summarises his gradings and places the candidates in rank order for each factor.
3 Through discussion and revision the assessors reach a final ranking on each factor.
4 Each candidate is now reviewed as a complete man with all positive and negative attributes considered by the asking of the following questions:
   How well will he do the job and fit in with his colleagues?
   To what extent can the company meet his personal requirements?
   How well does he fit into the company's long term manpower plan?
5 The final decision is now made.

All candidates should be advised of the decision and the statistical data collated and stored for future reference.

3.5.2 INDUCTION

It is important that every new employee is smoothly and gradually introduced to both the company and to his particular task. This process is known as induction and can commence immediately after his acceptance of the position.

1 Check final references (if remaining).
2 Arrange medical examination (if not already carried out).
3 Send moving into the area information if required.
4 Arrange introductions to bankers, schools, estate agents etc.
5 Check availability of equipment.

At about 14 days before the new entrant takes up his post:
1 Send joining information, required documents, instructions as to where and when and to whom to report.
2 Ensure department or site is prepared for his arrival.
3 Prepare an induction programme as necessary.
4 Inform all those with whom he will have contact of his arrival.

On the morning of his arrival
1 Make arrangements for his receipt.
2 Collect and issue necessary documentation.
3 Carry out all other required administrative procedures.
4 Give all vital information (not too much).
5 Issue necessary equipment.
6 Introduce to department or site.

3.5.3 FOLLOW-UP

During the induction programme and in the early days of job performance frequent monitoring of the new entrant's progress will ensure early identification of difficulties. By answering the following questions feedback is also provided regarding employment policy generally:

1 How successful are the job analyses, are they meaningful to the candidates and accurate in their predictions?
2 How successful is recruitment advertising, what is the cost and what is the response?
3 How satisfactory are the application forms?
4 How successful are the assessors' judgements when measured against actual performance, to what degree is their rating correct?

BIBLIOGRAPHY

The Chartered Institute of Building. *Occasional Paper No. 6 Personnel management,* M.J. Downham, CIOB Ascot
Boyce Martin, J, *Essentials of management: personnel management,* MacDonald and Evans (1977)
Plumbley, P and Williams, R., *The person for the job,* BBC Publications (1972)
Thomason, G.F., *A textbook of personnel management,* 3rd ed, The Institute of Personnel Management (1978)

EXERCISES

RECALL QUESTIONS

Identify eight operational areas of personnel management

(a) . . . .   (e) . . . .
(b) . . . .   (f) . . . .
(c) . . . .   (g) . . . .
(d) . . . .   (h) . . . .

2 List the aims of manpower planning

(a) . . . .
(b) . . . .
(c) . . . .

3 By completing the following diagram indicate the stages of the manpower planning cycle

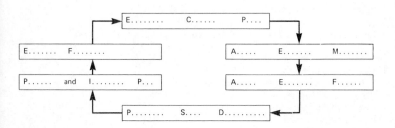

4 Complete the headings under which a Job Description may be written
   (a) J . . T . . . . and G . . . .   (e) W . . . . . . C . . . .
   (b) L . . . .   (f) T . . . .
   (c) D . . . . .   (g) R . . . .
   (d) R . . . .

5 Identify the headings which appear on a personnel specification
   Vertical Headings
   (a) . . . . (b) . . . . (c) . . . .
   Horizontal headings
   (d) . . . .   (g) . . . .
   (e) . . . .   (h) . . . .
   (f) . . . .   (i) . . . .

QUESTIONS REQUIRING SHORT ESSAY ANSWERS (15–20 minutes)

1 Describe four applicable objective tests which might be useful within the selection process.

2 Identify the difficulties faced by a building contractor attempting to prepare a formal manpower plan for the coming five years.

3 Under the headings; planning, conduct and decision explain how selection interviews can be made more effective.

4 List the information which may be kept on an employee record card and against each item state the reasons for its retention

5 Indicate the action you would take to ensure the following vacancies within a contractor's organisation are successfully filled:
   (a) Bricklayer.   (c) Site manager.
   (b) Trainee quantity surveyor.   (d) Managing director.

QUESTIONS REQUIRING LONGER ESSAY ANSWERS (30–45 minutes)

1 Explain in detail each of the six stages of manpower planning.

2 Using your own job, or any other about which you have detailed knowledge, write both a job description and personnel specification. From the latter prepare a block quarter-page advertisement for inclusion in the trade press.

3 Describe the seven point plan and explain how it might be used to improve selection procedures.

4 Prepare a detailed induction programme for a buyer newly appointed to the staff of a construction company.

5 Comment on the opinion that selection interviews are little better than useless in the selection of personnel.

4 Industrial relations

4.1 Historical development	4.1.1 The organisation of the building industry
	4.1.2 The Building Trades' Unions
	4.1.3 Building Trades Employers' Associations
	4.1.4 Joint negotiations

4.2 The Union of Construction, Allied Trades and Technicians	4.2.1 Responsibilities of trade unions
	4.2.2 The objectives of UCATT
	4.2.3 The organisation of UCATT

4.3 The National Federation of Building Trade Employers	4.3.1 The responsibilities of employers' associations
	4.3.2 The objectives of NFBTE
	4.3.3 The organisation of NFBTE

4.4 The National Joint Council for the Building Industry	4.4.1 Memorandum of agreement
	4.4.2 The Rules
	4.4.3 The structure

| 4.5 On-site industrial relations | 4.5.1 Trade Union recognition and procedures |
| | 4.5.2 Site procedure agreements |

4.1 HISTORICAL DEVELOPMENT

The right of workpeople to freely negotiate their conditions of employment is often considered to be a part of a long democratic tradition in this country. However, modern trade unions, as a reflection of that freedom, have difficulty in producing a written record prior to the beginning of the nineteenth century.

The Industrial Revolution is generally considered to be the cradle of trade unionism. Large scale production brought large numbers of workers together who, with little hope of advancement took the opportunity to form organisations to voice their grievances. Before this date industrial processes were carried out in small units. However certain sections of the building industry, those concerned with the erection of castles, fortifications, palaces and cathedrals, met the two preconditions long before 1750. There is evidence that before that date building workers attempted to impose conditions upon their employers.

4.1.1 THE ORGANISATION OF THE BUILDING INDUSTRY

The need for the rapid erection of fortifications in the 11th to the 14th centuries required a considerable number of craftsmen, mainly carpenters and masons, to be brought and lodged on the site until work was complete. As communities developed around these castles for their protection, the 'lodges' became permanent organisations and a formal system of apprenticeship developed. In a 7-year period the skills of the trade were passed from master to apprentice. When this period was completed it was necessary for some of the newly trained craftsmen to leave the lodge and search for work further afield and hence the three tier system of master, journeyman and apprentice became established.

The development of larger towns during the 14th century saw the lodges develop into permanent powerful institutions called Guilds. These Guilds occupied a monopolistic position within a town particularly in times of shortage of work. The upper positions were held by master craftsmen who, through the development of stringent financial rules, consolidated their position by excluding journeymen without capital from becoming masters. Because of this privileged position masters were approached by building owners to carry out work partly with their own employed journeymen and apprentices and partly by subcontracting work to other journeymen who had, by this time, formed craft associations having been excluded from the original Guilds.

Under the Guild system the slow pace of work allowed building details to be worked out by the master and the client as the work progressed. By the 17th century the increased pace of work, technical innovation and the increasing awareness of design required the interposition of the architect between the client and the master craftsmen. As most of the owners were of the nobility architects tended to be drawn from the ranks of educated gentlemen rather than the artisan class and hence the 'profession' was established within a formal training system.

By the 18th century the pattern shown in *Fig 4.1* had developed with the use of a measurer by the master to conduct financial negotiations with the architect. This resulted in the employment of a corresponding measurer by the client.

In the 19th century, as foundations were invariably of brick, the master bricklayer was generally first on the site. It was he therefore who began to incorporate the other major trades into his organisation and become established as the main contractor who sublet work to other trades. This organisation is illustrated in *Fig 4.2*.

It can be seen within this development that the major features of the traditional contracting system described in chapter one had by this time developed. The architect

59

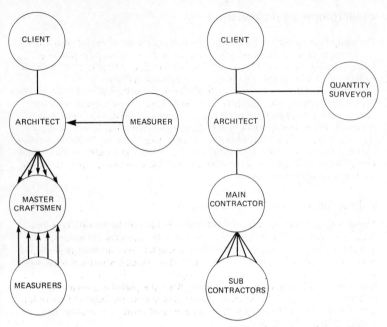

Fig 4.1 Organisation of the building industry in the 18th Century

Fig 4.2 Organisation of the building industry in the 19th Century

and quantity surveyor contracted on a fee basis to the client. The main contractor employed his own technical and operative staff and engaged subcontractors directly.

The Guilds now formed less formal groupings which were to develop into employers trade associations. The craft associations, formed by journeymen, concentrated on the provision of benefits and eventually became the friendly societies which in turn developed into craft based unions.

4.1.2 THE BUILDING TRADES' UNIONS

After years of repression the repeal of the Combinations Act in 1824 provided the trigger for the establishment of groups of workers who were to form the basis of modern trade unionism. The building crafts with their strong traditions rejected the pleas of Robert Owen, the father of British socialism, to join the Grand National Consolidated Trades Union and formed their own craft based unions instead. The General Union of Carpenters and Joiners was formed in 1827 followed in Manchester in 1829 by the Friendly Society of Bricklayers. A third union, more national in character, was the Operative Stonemasons Society which recruited 4000 members from Carlisle to Plymouth.

The development of the single contractor created a threat to the small local master craftsmen who, in a loose federation with the craft unions, formed the Operative Builders Union, to ensure that 'no new building should be erected by contract with one person'. However following a series of abortive demands the federation dissolved when the masters attempted to exclude the journeyman from membership of the General Union of Building Trades by the signing of a declaration which became known as 'The Document'. At the instigation of Robert Owen the journeymen attempted to by-pass the masters by contracting directly to building owners through the National Building Guild. However because of fragmented organisation and lack of funds this early cooperative venture failed and many craft unions with it.

The result of this catastrophe was the realisation that a high degree of central control was necessary and the appointment of full time General Secretaries and improvements in communications allowed many autonomous branches to merge forming national associations. The 1860s saw the emergence of the Associated Society of Carpenters and Joiners, the National Association of Operative Plasterers and the United Plumbers Association. Whilst most developments to date concerned the crafts in 1889 the United Builders Labourers Union was formed.

The concept of a single union for the building industry was again raised in 1914 with the formation of the Building Workers Union, but, because of resistance from the established unions and the outbreak of the first world war, this proved short lived. The merits of such a union were however recognised and the National Federation of Building Trade Operatives was founded as a compromise solution. This body was a tightly knit federation of the existing trade unions and it was to play a most significant role in joint negotiations for the next 50 years. The 1920s and 1930s saw the beginnings of amalgamations between trade unions which laid the foundations for the post war period. The Amalgamated Society of Woodworkers (ASW) was established and the bricklayers and masons formed the Amalgamated Union of Building Trade Workers (AUBTW).

Union membership fluctuated widely responding to the level of activity in the industry but the development in the 50s and 60s of labour only subcontracting resulted in a serious decline in membership to a point that functioning was hindered through lack of finance. The cost of affiliation to the NFBTO was an increasing burden forcing many unions to reduce their affiliated membership and the National Association of Operative Plasterers withdrew to merge with the Transport and General Workers Union forming a building craft section. A general deterioration in relationships and the financial pressures resulted in the abandonment of any centralised union machinery thus leaving a vacuum in the negotiations with the employers federation.

In 1970 the Amalgamated Society of Painters and Decorators and the Association of Building Technicians both merged with the ASW who then commenced merger discussions with the AUBTW, the other dominant union. (The plumbers and electricians had by this time joined forces but allied themselves to engineering rather than construction.)

By December 1981 ASW and AUBTW had reached agreement on a new set of rules and organisation procedures and the Union of Construction, Allied Trades and Technicians (UCATT) was established.

In the civil engineering sector craft operatives are generally also members of UCATT, whilst the TGWU has a substantial membership particularly in open cast coal mining, and the GMWU is strong in gas distribution.

Table 4.1 on page 62 indicates the current representation of building industry workers.

4.1.3 BUILDING TRADES EMPLOYERS' ASSOCIATIONS

Whilst building operatives had much to gain from collective action, their employers were isolated by the often bitter competition for work. The history of building trades

Table 4.1 Major representation of the Building Trade Unions

Union	Representing
Union of Construction, Allied Trades and Technicians (UCATT)	Bricklayers Joiners Painters Stonemasons
Transport and General Workers Union (TGWU)	Mechanical plant operatives Plasterers Scaffolders Roof slaters and tilers
General and Municipal Workers Union (GMWU)	Non craft operatives (also represented by UCATT and TGWU)
Furniture, Timber and Allied Trades Union (FTAT)	Cabinet makers French polishers Woodcutting machinists
Electrical, Electronic, Telecommunication and Plumbing Union (EETPU)	Electricians Plumbers

employers' associations is, therefore, far shorter than that of the trade unions. Where associations sprang into existence it was sporadic and generally out of self interest in response to specific union pressure in a local area.

In 1858 the National Association of Master Builders was established but, despite the title, its strength was concentrated in a few of the larger towns of the north and midlands. The Association was, however, shortlived for it overplayed its hand in attempting to reintroduce payment by the hour instead of the accepted daily rate. Within a mere 14 days of the ensuing dispute solidarity had crumbled with just a few employers continuing the fight.

In 1861 the London employers proved much stronger in their organisation and, following a 6-week lockout, they succeeded where the northern employers failed. This success resulted in the formation in 1865 of the General Builders Association but strong regional affiliations were maintained with the establishment of the London Master Builders Association and the Liverpool Master Builders Association.

Eventually in 1878 the struggle for national unity succeeded with the formation of the National Association of Master Builders of Great Britain which, in 1899, adopted their present title of the National Federation of Building Trade Employers (NFBTE). In 1906 the first full time General Secretary was appointed and, upon the establishment of the NFBTO in 1919, joint negotiations took place representing the largest affiliations of operatives and employers in the building industry.

It is apparent from the history that one of the major obstacles to unity has been the range of sizes of organisations from the small local firm to large national and lately international contractors. Even today it is felt by many that no single institution can represent the whole spectrum of sizes and interests and, although this is contested by

Table 4.2 Major representation of Building and Civil Engineering Employers Associations

Employers Association	*Representing*	
National Federation of Building Trade Employers (NFBTE)	12000	Building and Sub-contractors of all sizes
Scottish Builders Employers Federation (SBEF)	1800	Building and Sub-contractors of all sizes
Federation of Master Builders (FMB)	20000	Small/medium contractors and self-employed
Federation of Civil Engineering Contractors (FCEC)	600	Large/medium civil engineering contractors
Federation of Specialist Sub-Contractors (FASS)		Sub-contractors of all sizes

some, the Federation of Master Builders has constantly endeavoured to represent the interests of the small builder to central authority.

Other employers associations have developed to cater for the specialist interests; the House Builders Federation, the National Federation of Painting and Decorating Contractors and the National Federation of Roofing Contractors, for example.

Table 4.2 summarises the current situation.

4.1.4 JOINT NEGOTIATIONS

It has been demonstrated that local affiliations of labour and employers were the earliest seats of joint negotiations. A number of experiments around the turn of the century led to the establishment of 'local joint councils'. In 1908 these local initiatives led to the establishment of the National Board of Conciliation which considered wages and conditions in the industry. After the 1914—18 war wider issues were considered by the Industrial Council for the Building Industry but in 1921 the National Wages and Conditions Council for the Building Industry was established and quickly gained approval for a national agreement. In 1926 it was renamed the National Joint Council for the Building Industry (NJCBI) and when, in 1932, it absorbed the old National Board of Conciliation it became the institution which largely operates today. In 1930 the Scottish employers had established the Scottish Joint Council operating an agreement similar to the English, but this separation ended in 1964 when the two councils merged.

Up to 1972, eleven separate unions represented the employees on the NJCBI. However the amalgamations which led to the formation of UCATT and to the merger of the plumbers with the electricians reduced this number to 4: UCATT, TGWU, G & MWU and FTATT. Although the constitution of this body has changed in the past, and will change in the future its current constitution is:

Employers	NFBTE	15 (incl. SBEF)
	NFRC	1

Employees	UCATT	8
	TGWU	4
	G & MWU	2
	FTAT	2
Total		32

The growing pressure and complexity of industrial relations and collective bargaining resulted in the establishment, in 1974, of the Building and Civil Engineering Joint

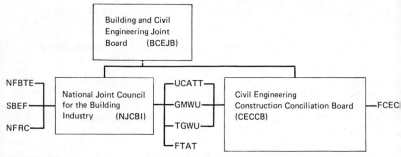

Fig 4.3 Joint negotiating machinery for the construction industry

Board (BCEJB) which endeavours to promulgate agreements acceptable to both the NJCBI and the Civil Engineering Construction Conciliation Board. The current situation is illustrated in *Fig 4.3*.

Reference has been made to attempts by the FMB to gain representation on the NJCBI and when a formula could not be found in the late 70s a separate agreement

Table 4.3 Joint negotiating machinery for specialist interests

Joint machinery	Employers	Unions
Joint Industrial Board for the Electrical Contracting Industry	Electrical Contractors Association	EETPU
Joint Industrial Board for Plumbing and Mechanical Engineering Services	National Association of Plumbing, Heating and Mechanical Services Contractors	EETPU
Demolition Industry Conciliation Board	National Federation of Demolition Contractors Ltd.	GMWU TGWU UCATT
National Joint Council for the Thermal Insulation Contracting Industry	Thermal Insulation Contractors Association	GMWU TGWU
Plant Hire Working Rule Agreement	Contractors Plant Association	GMWU TGWU UCATT

was reached between the FMB and the TGWU establishing the Building and Allied Trades Joint Industrial Council (BATJIC) as an alternative joint negotiating body. BATJIC reached its first agreement in 1980 and has since established a wider set of agreements.

In addition to these two bodies there exists other groupings representing specialist interests within the construction industry. These bodies are identified in *Table 4.3*.

4.2 THE UNION OF CONSTRUCTION, ALLIED TRADES AND TECHNICIANS (UCATT)

4.2.1 RESPONSIBILITIES OF TRADE UNIONS

The Industrial Relations Code of Practice, retained in parts despite the repeal of The Industrial Relations Act which instigated its publication, sets out the responsibilities of trade unions in the following terms:

'The trade unions principal aim is to promote its members' interests but it also shares with management the responsibility for good industrial relations.'

In order to achieve these twin objectives it must be prepared to establish and maintain effective arrangements for joint negotiations with employers, and other trade unions, relative to the settlement of disputes and differences. It must also take reasonable steps to ensure that its officials are adequately trained in the understanding of their responsibilities and that all members observe agreements and recognised procedures. The trade union must establish an efficient organisation structure to ensure effective communication between the membership, its elected officers and full-time officials. Members on their part must provide the union with sufficient resources and the authority needed to carry out its functions.

4.2.2 THE OBJECTIVES OF UCATT

The general aim of the trade union is 'to promote the social and economic advancement of the members and of the workers generally'. In order to achieve this aim the following specific objectives have been identified:
1 To agree and secure observance of equitable rates of wages and conditions at national and local levels.
2 To represent the views of its members to employers and employers federations, to government departments, public and professional bodies and the wider trade union movement.
3 To seek an improvement in the status of its members.
·4 To encourage and promote proper systems of training.
5 To encourage and promote proper systems of employment.
6 To provide expertise on behalf of members involved in lawful disputes and provide representation at industrial tribunals.
7 To develop the knowledge and expertise of full-time and part-time officials.
8 To further the interests of trade unionism generally.

4.2.3 THE ORGANISATION OF UCATT

In 1971 the Union of Construction, Allied Trades and Technicians was formed through the amalgamation of existing unions to become the largest single union in the construction industry, representing in 1978, 325,245 workers, establishing UCATT as the 10th largest union. The democratic process requires that each worker is able to influence union policy and the organisation exists to accommodate this aim.

In 1979 there were 1600 branches varying in size and constitution. Some branches are representative of many crafts assembled in one place whilst others represent one craft with members drawn from a number of sites in one area.

Branches meet fortnightly or monthly by agreement and business is contracted in accordance with procedural rules with voting generally on a show of hands. The branches elect members to twelve Regional Councils which maintain and protect the general and working conditions of employment of all members within that region. Also elected by the branches is the seven member Executive Council which administers the rules of the union; and the ten-member General Council, the main appeals body.

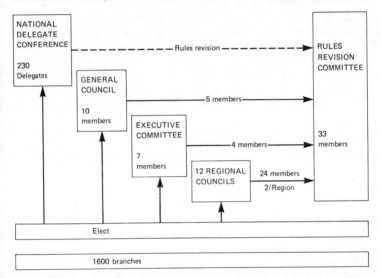

Fig 4.4 The organisation of The Union of Construction, Allied Trades and Technicians

Once every two years 230 delegates are elected to the National Delegate Conference which is the policy making body of the Union. Provision is made for the initiation of policy between conferences and the Executive Council is empowered to seek the views of the membership by ballot where there is insufficient guidance within the constitution and rules of the union. Should exception be taken to any decision of the Executive Council it may be set aside upon appeal to the General Council. Where any rule requires amendment or a new one introduced the Rules Revision Committee must investigate. This body comprises thirty-three members; two from each region, four from the Executive Committee and five from the General Council. *Fig 4.4* summarises the elected policy and decision making bodies within the union.

4.3 THE NATIONAL FEDERATION OF BUILDING TRADES EMPLOYERS (NFBTE)

4.3.1 THE RESPONSIBILITIES OF EMPLOYERS ASSOCIATIONS

The Industrial Relations Code of Practice charges employers associations with certain responsibilities in the same manner as it did for trade unions. The prime aim of an employers association is to promote those interests of its members which can best be served by inter-firm cooperation.

They should therefore maintain joint negotiations with trade unions to establish terms and conditions of employment and encourage members to observe agreements. They should also provide an industrial relations advisory service for their members through which information may be disseminated, trends identified and advice proffered.

Individual members should on their part provide the association with the necessary authority and resources in order that its stated function may be fulfilled.

4.3.2 THE OBJECTIVES OF THE NFBTE

Acting as both collective bargainer and trade association the objectives cover a wide area but the following have been identified as being amongst the most important:
1 To represent members in their individual and collective capacities and to protect them against attack.
2 To secure for members the greatest possible freedom in the conduct of their business.
3 To develop and maintain friendly relations between members.
4 To regulate relations between employers and workers and to promote regular procedures in regard to the negotiation of wages, hours and other conditions of employment, the adjustment of any differences arising therefrom, and the prevention of strikes.
5 To secure fair and equitable conditions of tendering, bills of quantities, specifications, conditions of contract and the supply of materials.
6 To examine all legislative measures which may affect the building industry and take all necessary measures.
7 To maintain a high standard of conduct and efficiency and to combat unfair practices.

4.3.3 THE ORGANISATION OF NFBTE

The National Federation of Building Trade Employers geographically covers England and Wales and has about 12 000 firms of all sizes in membership. Some are in direct membership and some are in affiliated specialist associations such as:

Scottish Building Employers Federation.

National Federation of Plastering Contractors.

House Builders Federation.

England and Wales are divided into ten regions which are subdivided into 173 local associations and ten local areas as shown in *Fig 4.5*.

The rights and status of the membership are protected by the Constitution and Rules, and control over policy and administration is exercised by the National Council. Of the ninety-five members, sixty-three are regional representatives, fourteen are members of the affiliated bodies and eighteen are ex-officio or coopted members.

National	NATIONAL FEDERATION OF BUILDING TRADE EMPLOYERS									
Regional	EASTERN	LIVERPOOL	LONDON	MIDLAND	NORTHERN COUNTIES	NORTH WESTERN	SOUTHERN	S. WALES	S. WEST	YORKS
Regional office	Cambridge	Liverpool	London	Birmingham	Durham	Manchester	Horsham	Cardiff	Bristol	York
Local associations	21	–	5 areas	23	5 areas	40	36	14	13	26

Fig 4.5 The organisation of The National Federation of Building Trade Employers

Officers are elected at the Annual General Meeting and the administration is conducted by full-time staff who draw upon the expertise of specialists in, for example, industrial relations, the law, education and training. The head office staff are supplemented by full-time personnel in the affiliated regional federations and local associations.

In order that the federation may represent the interests of their diverse membership each member is allocated to one of the following sections:

1 National Contractors Section.

2 Smaller Builders Section.

3 System Builders Section.

To accommodate geographical variations Regions are not dominated by the National Council completely and may, within certain limitations, vary the Rules and Constitution at their discretion. Nineteen standing sub-committees exist to prevent National Council being overwhelmed by business, and these committees include:

Public relations and membership	Plumbers
Finance	Training
Management	Taxation
Smaller builders	Insurance
Sub-contractors	

The interests of its members are represented in direct consultation with the government through such bodies as:

Economic Development Committee for Building.

Building Research Establishment.

British Standards Institution.

Health and Safety at Work Commission.

and also on many institutions and professional bodies such as:

The Chartered Institute of Building.

Construction Industry Training Board.

Joint Contracts Tribunal.

4.4 THE NATIONAL JOINT COUNCIL FOR THE BUILDING INDUSTRY (NJCBI)

The National Joint Council for the Building Industry has been introduced by this chapter as the dominant machinery during this century for the conduct of joint consultation and negotiation between employers and unions. Whilst significant changes have recently taken place this body remains influential in a large sector of the construction industry.

Control of the NJCBI is exercised through the constitution, rules and regulations which is a three part agreement issued after revision in 1972.

Part I Memorandum of Agreement.

Part II Rules.

Part III Regulations controlling submissions by Regional Councils.

4.4.1 MEMORANDUM OF AGREEMENT

The Memorandum sets out the operating areas of the NJCBI and indicates ways in which the specific objectives may be met:

1 The regulation of wage rates;

2 Terms and conditions of employment.

3 Schemes of training.

4 Settlement of disputes and differences.
5 Consideration of industrial or economic questions.
6 Joint representations to government and other bodies.

The mechanism for the achievement of the first four objectives is the publication of an agreed set of terms and conditions called the National Working Rules for the Building Industry.

4.4.2 THE RULES

The rules of procedure require the council to meet a minimum of four times a year with the Annual Meeting held in April. Items for decision are put to a show of hands and in order for a motion to be carried majorities are required on both sides of the table. In other words both employers and employees must be in favour. The Officers of the Council include a Chairman and a Vice-Chairman (by custom an employer member and an employee member respectively) who are elected from the membership. Each side of the Council elect an Honorary Joint Secretary who, with a full time staff under the control of a Clerk to the Council, deal with all administrative matters.

4.4.3 THE STRUCTURE

In order to carry out its functions the Council appoints a series of standing committees, for example a Management Committee, a National Conciliation Panel and a National Joint Training Commission. It may also establish other committees as required to deal with referred matters but such committees have no decision making power and must present their findings to the Council.

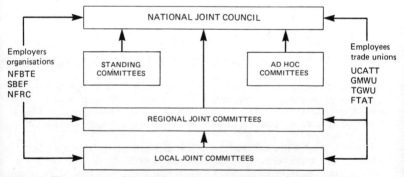

Fig 4.6 The structure of the National Joint Council for the Building Industry

The National Joint Council delegate certain authority to 9 Regional Committees and to 1 serving Liverpool and District. These Regional Committees are responsible for:
1 The operation of the working rule.
2 The initial proposals for regional amendments to the working rules.
3 The dealing with disputes and differences.

Where appropriate the Regional Committee may appoint Local Joint Committees who are responsible for:
1 The operation of the working rules in their area.

2 The regulation of overtime working.

3 The dealing with disputes and differences.

This arrangement is illustrated in *Fig 4.6.*

4.5 ON-SITE INDUSTRIAL RELATIONS

Whilst most of the operatives employed on building sites are subject to working rules of one kind or another, there is still an element of discretion and interpretation which is to be resolved on site between representatives of site management and the trade unions. In order for such negotiations to be concluded successfully both parties will require skills in the art of compromise. The identification of a procedure most likely to achieve a satisfactory outcome must be identified. Such a procedure is summarised in *Fig 4.7.*

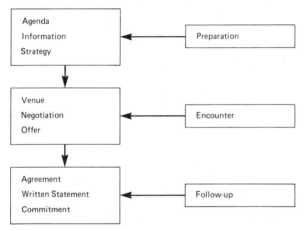

Fig 4.7 Phases of a negotiation

4.5.1 TRADE UNION RECOGNITION AND PROCEDURES

For those operatives under the jurisdiction of the NJCBI, National Working Rule 25 under the heading 'Trade Union Recognition and Procedure' sets out agreed procedures for the conduct of on-site negotiations. The privileges and rights given to trade union representatives only apply to those unions who are parties to the agreement.

Full time trade union officers Any full time officer of a trade union, whose name has been submitted to the Regional Employers' Secretary, may, by arrangement, have access to sites, shops and jobs to ensure observance of the working rules.

Accreditation of site representatives When an operative has worked for 4 weeks on a site he is eligible for selection as a representative and employers may not unreasonably withold recognition following written confirmation of the Union. Normally only one representative for each trade or union need be recognised.

The duties and functions of such a representative are as follows:
1. To represent the members.
2. To ensure proper observance of the rules especially where related to disputes and difference.
3. To recruit to appropriate unions.

Convenor stewards When a number of representatives are recognised on one site, one of them may be accredited as convenor steward and once again, when informed in writing, the employer may not unreasonably withold recognition. His duties will be:
1 To represent operatives when more than one trade or union is involved.
2 To assist individual representatives in their duties.
3 To act as a representative of the Operatives' Regional Secretary.
4 To ensure observation of the Working Rule Agreement particularly in relation to dispute and differences.

In order to carry out such duties the steward should be allowed access to a meetings room, a telephone and a notice board; he should be allowed to leave his place of work, after seeking permission, to conduct urgent relevant business and he should be allowed reasonable time off work to attend recognised training courses in industrial relations.

Site representatives committee The convenor steward should be chairman of any Site Representatives' Committee established to assist in the proper conduct of industrial relations. Such a committee should number no more than 7 members elected from the representatives, and their names should be transmitted to the employer through the Operatives' Regional Secretary. Proper minutes of the committee should be kept and a full time trade union officer may be nominated an ex-officio member. The committee must seek approval for any meeting held during working hours and may take no action which is in variance to the Working Rule Agreement.

4.5.2 SITE PROCEDURE AGREEMENTS

Whether a site is under the jurisdiction of the NJCBI or not a site agreement may be sought by either party in an effort to establish a negotiated base for the conduct of industrial relations on site. The objectives of such an agreement are to remove difficulties before they disrupt progress on site and to specify precise site conditions. A typical procedure may start:

A PROCEDURE AGREEMENT on the conduct of
SITE INDUSTRIAL RELATIONS between
XYZ CONSTRUCTION PLC (hereinafter called the Company)
and
TRANSPORT AND GENERAL WORKERS UNION
GENERAL AND MUNICIPAL WORKERS UNION
UNION OF CONSTRUCTION, ALLIED TRADES AND TECHNICIANS
(hereinafter called the Unions)
for work to be carried out by the Company on the
URBAN CENTRE RENEWAL PROJECT LONCHESTER

1 Objectives
  To establish and maintain good industrial relations on site through the cooperation of all parties.
  To ensure the proper resolution of conflict without the disruption of work.
  To improve the efficiency and productivity and to reward such improvement with increased earnings.

2 Scope

The agreement shall apply to all operatives directly employed by the Company and all sub-contractors employed by them.

Then will follow a variety of detail covering all required areas which will probably include:

3 The appointment and recognition of shop stewards and convenor stewards.
4 Meetings of operatives.
5 Individual and joint meetings.
6 General conditions of employment.
7 Working hours.
8 Bonus.
9 Grievance procedure.
10 Safety.
11 Discipline.
12 Amenities.

and will probably finish with

13 Amendments to this Agreement. Any required amendment to this document must be agreed by all the signatories after a minimum of 2 months written notice.
14 In signing this Agreement the signatories recognise that, although it will be honoured by all parties, it has no legal enforceability.

BIBLIOGRAPHY

Advisory, Conciliation and Arbitration Service, *Industrial relations handbook,* HMSO (1980)

Higgins, G. and Jessop, N., *Communications in the building industry,* Tavistock Publications (1965)

Hilton, W.S., *Industrial relations in construction,* Pergamon (1968)

National Joint Council for the Building Industry. *Working Rule Agreement* (1982 edition)

Wood, L.W., *A Union to build : The story of UCATT,* Lawrence and Wishart (1979)

EXERCISES

RECALL QUESTIONS

1 Identify the following trade unions:
   (a) EETPU; (b) FTAT; (c) GMWU; (d) TGWU; (e) UCATT

2 List the three stages of negotiations:
   (a) P . . . . ; (b) E . . . . ; (c) F . . . .

3 Complete the diagram representing the structure of the NJCBI

N . . . .   J . . . .   C . . . . . .
R . . . .   J . . . .   C . . . .
L . . . .   J . . . .   C . . . .

4   Identify the following employers' associations
    (a) FASS; (b) FCEC; (c) FMB; (d) NFBTE; (e) SBEF

5   List four major areas that might be covered in an on-site procedural agreement
    (a) ....
    (b) ....
    (c) ....
    (d) ....

QUESTIONS REQUIRING SHORT ESSAY ANSWERS (15—20 minutes)

1   Briefly outline the development of
    (a) Employers' Associations; (b) Trade Unions
    within the construction industry.

2   Identify the major objectives of a trade union.

3   As a convenor steward on a large building site you are about to undertake
    negotiations for an agreement. Identify your aims for amenities in excess of the
    minimum standards and prepare your substantiating argument.

4   Describe the functions and responsibilities of a trade union site representative.

5   List the major objectives of an employers' association.

QUESTIONS REQUIRING LONGER ESSAY ANSWERS (30—45 minutes)

1   'Trade unions have too much power and too few responsibilities' Discuss.

2   Describe the development of the negotiating machinery within the building
    industry in the 20th century.

3   'Because of the differences between construction companies no single association
    can truly represent them all' Discuss.

4   Analyse in detail the probable content of a site procedural agreement between a
    building company and the trade unions.

5   As convenor steward you are Chairman of the Site Representatives Committee
    about to be constituted. Prepare a report to explain your objectives to site manage-
    ment and list your constitution and operating procedures.

5 Industrial relations legislation and Codes of Practice

5.1 Employment legislation	5.1.1 Legislation on industrial relations 5.1.2 Employee rights

5.2 National Mechanisms of Industrial Relations	5.2.1 Certification Officer 5.2.2 The Advisory Conciliation and Arbitration Service 5.2.3 The Central Arbitration Committee 5.2.4 Industrial tribunals 5.2.5 Employment Appeal Tribunal 5.2.6 Codes of practice 5.2.7 The Industrial Relations Code of Practice

5.3 Codes of Practice issued by ACAS	5.3.1 CoP 1 Disciplinary practice and procedures in employment 5.3.2 CoP 2 Disclosure of information to Trade Unions for collective bargaining purposes 5.3.3 CoP 3 Time off for Trade Union duties and activities

5.4 Codes of Practice issued by the Secretary of State for Employment	5.4.1 CoP Closed shop agreements and arrangements 5.4.2 CoP Picketing 5.4.3 Draft CoP The elimination of racial discrimination and the promotion of equality of opportunity in employment

5.1 EMPLOYMENT LEGISLATION

It has been suggested that employment law developed through history by conflict between the two sources of legislative authority; Parliament and the Courts. Up to the middle of the 19th century the Establishment were opposed to trade unionism and used common law relating to conspiracy and restraint of trade to combat its development. The transportation of the so called Tolpuddle Martyrs proved a turning point leading as it did to the Trade Union Act of 1871 which provided certain protection for the individual.

The Taff Vale judgement of 1901 however, by awarding damages of £40 000 against a trade union, established their liability in civil actions. The large Liberal majority in 1906 enabled the passing of the Trade Disputes Act of the same year prohibiting legal actions in tort (civil wrongs) against persons acting 'in contemplation or furtherance of a trade dispute'. Subsequent historic events and economic difficulties restricted the power of trade unions and policies endeavoured to produce an atmosphere of voluntary agreement in which industry could prosper.

In the 50s and 60s, with the country moving towards a 'full' employment era, trade union strength and power increased and fear of politically motivated disruption of employment created torts relating to the inducement to break employment and commercial contracts. The Trade Disputes Act 1965 provided some additional protections but, following the report of a Royal Commission in 1968, both the Labour Government (*In Place of Strife*) and the Conservative Opposition (*Fair Deal at Work*) served notice of things to come.

It was the Conservatives who started a decade of rapidly changing legislation with the passing in 1971 of the Industrial Relations Act. In the long term this Act failed probably because:

1 It attempted to replace voluntary agreements binding in honour with a highly centralised legislative framework.
2 It did not get the active support of management.
3 It generated fierce trade union resistance.

Two established elements did however survive; firstly the Code of Industrial Relations Practice which set out a basis of conduct in areas where no enforceable standards existed, and secondly the concept of unfair dismissal. The change of the Government in 1974 resulted in the immediate repeal of the Industrial Relations Act and all legislation since that date, although revised, still remains on the statute book.

In reading this chapter it must be recognised that any summary must inevitably suffer through omission and simplification. Only the official wording of the Act and subsequent interpretation by the Courts are authoritative. Also it is inevitable that changes will occur and the reader is therefore advised to consult the latest references and seek specialist advise on current legislation and its interpretation. Of particular importance at the moment is the progress of the 1982 Employment Bill through the Houses of Parliament.

5.1.1 LEGISLATION ON INDUSTRIAL RELATIONS

Trade Union and Labour Relations Act 1974 as amended by the Employment Act 1980

1 Sets out conditions under which 'no strike' clauses become legally binding.
2 Defines a trade union.
3 Establishes right to terminate trade union membership.

4   Defines employers' associations.
5   Defines Trade Dispute and Secondary Action.
6   Originates a list of trade unions and employers associations.
7   Establishes conditions for legal picketing.
8   Retains the Code of Industrial Relations Practice previously approved by the Industrial Relations Act 1971 until such times as it is revised.

Employment Protection Act 1975 as amended by the Employment Act 1980
1 Establishes the Advisory, Conciliation and Arbitration Service and Certification Officer.
2 Defines 'Recognised' trade unions.
3 Establishes rights in relation to the disclosure of information.
4 Sets out procedures for the handling of redundancy.

Employment Act 1980
1 Provides financial reimbursement for certain ballots and establishes a right of use of employers premises for the carrying out of ballots.
2 Gives the Secretary of State the right to issue Codes of Practice (in addition to ACAS).

Employment Bill 1982
If enacted as planned this Bill will:
1 Enable the Government to compensate people dismissed under closed shop regulations.
2 Increase the rights of non-union employees.
3 Promote reviews of existing closed shop agreements.
4 Make trade unions liable for unlawful industrial action.
5 Restrict lawful trade disputes to those between workers and their own employer.

5.1.2 EMPLOYEE RIGHTS

Employment Protection (Consolidation) Act 1978 as amended by the Employment Act 1980
1 Establishes the right of each individual to receive written conditions of employment and an itemized pay statement.
2 Sets out weekly pay guarantees.
3 Protects individuals who are suspended on statutory medical grounds.
4 Establishes the right of an individual to elect to join, or not to join, a trade union.
5 Regulates the allowance for time off work for specified activities.
6 Affords protection to female employees in relation to maternity.
7 Sets out the rights to notice of termination of employment and establishes minimum periods.
8 Establishes procedures and remedies in relation to unfair dismissal.
9 Defines dismissal and establishes procedures and scales of payment in cases of redundancy.

Race Relations Act 1976
1 Defines permitted and unlawful discrimination and establishes compensation.
2 Establishes the Commission for Racial Equality to replace the Commission for Community Relations and the Race Relations Board
3 Provides for Race Relations Employment Advisors.

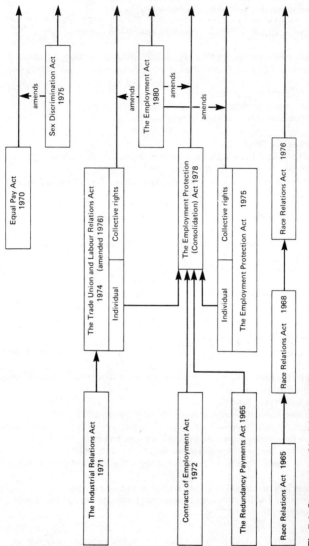

Fig 5.1 Summary of legislation affecting industrial relations to December 1981

Sex Discrimination Act 1975
1 Specifies acts of discrimination.
2 Confirms the duties of the Equal Opportunities Commission
3 Establishes remedies in cases of discrimination.
4 Amends areas of the Equal Pay Act 1970.

Equal Pay Act 1970
1 Establishes the right to equal pay in conditions of broadly equal work.
2 Provides remedies in cases of proven inequalities.
3 Established a 5-year phasing in period 1970–1975.
Figure 5.1 traces the development of industrial legislation from 1971 to 1981.

5.2 NATIONAL MECHANISMS OF INDUSTRIAL RELATIONS

5.2.1 CERTIFICATION OFFICER

The Certification Officer is an independent authority appointed by the Secretary of State for Employment who, under the EPA 1975, TULRA 1974 and EA 1980 is responsible for:
1 Maintaining a list of trade unions and employer associations.
2 Determining the 'independence' of trade unions.
3 Oversight of the audited returns of trade unions and employer associations.
4 Supervising the statutory requirements related to the establishment of political funds.
5 Refunding the costs of certain ballots.
6 Holding for public inspection the annual returns and rule books of trade unions and employer associations.

5.2.2 THE ADVISORY, CONCILIATION AND ARBITRATION SERVICE (ACAS)

The Advisory, Conciliation and Arbitration Service was established in 1976 under the EPA 1975 and is now charged with the following duties:
1 To offer conciliation and other assistance to help settle trade disputes.
2 To provide Conciliation Officers to promote the settlement of complaints made to industrial tribunals.
3 To refer disputed matters to independent arbitration or to the Central Arbitration Committeee.
4 To offer advice to employers, employer associations, workers and trade unions, relative to industrial relations, employment policy and general matters.
5 To enquire into any matter relating to industrial relations within a firm or industry.
6 To issue Codes of Practice relative to the promotion of good industrial relations.
ACAS is controlled by a Council comprising a full time Chairman and nine other members appointed by the Secretary of State for Employment, 3 after consultation with the CBI and 3 after consultation with the TUC. Apart from these appointments and the presentation of accounts the service is free from government interference. Advice on industrial relations is given through:
1 The answering of queries.
2 Short advisory visits.
3 In-depth investigations.

4 The organisation of seminars and courses.

ACAS attempts to solve disputes by conciliation, mediation or arbitration.

1 Conciliation procedures, though flexible, generally commence with the appointment of a mutually acceptable conciliator who has no powers other than reason and persuasion. Through joint or separate meetings he will attempt to bring together the two sides in agreement.

2 The role of the mediator requires that he bring original formal proposals to the two sides which may form the basis of an agreeable solution.

3 If both parties are willing and all other procedures have been exhausted, arbitration may take place. The main stages of ACAS arbitration are:
   (a) The appointment of an arbitrator or a board or arbitration.
   (b) The preparation and exchange of written statements.
   (c) An oral hearing in the presence of both parties.
   (d) The consideration and announcement of the award.

5.2.3 THE CENTRAL ARBITRATION COMMITTEE (CAC)

The Central Arbitration Committee was established at the same time as ACAS. The Secretary of State for Employment appoints a Chairman, Deputy Chairman and members who are drawn from both sides of industry. Hearings may result from references by ACAS, be specified in the final stages of the disputes procedure of an organisation or be the result of specific legislation.

Hearings are usually conducted before the Chairman and one member from each side of industry and the decisions are published in the form of an award which also includes the deliberations.

5.2.4 INDUSTRIAL TRIBUNALS

Whilst industrial tribunals were established under the National Insurance Act of 1911, it was the Industrial Training Act 1964 which increased the requirement and all subsequent legislation has included such hearings as part of appeals procedure. Tribunals are independent judicial bodies each having a legally qualified Chairman and two other members, one drawn from a panel of members recommended to the Secretary of State for Employment by employers and one from a similar panel recommended by trade unions.

Complaints are originally made in writing to the Secretary, responsible for all administration, who after a simple vetting procedure, asks the respondent (generally the employer) if he intends to resist the claim. If so copies of the relevant documents are forwarded to ACAS who will attempt reconciliation. If an appearance at a tribunal becomes inevitable it is the Chairman's duty to ensure that all available evidence is heard irrespective of the powers of expression of the applicant, who may or may not be represented by a legal representative or trade union official. Tribunals may be conducted in a flexible manner but the following may be considered standard procedure:

1 Opening statement by respondent.
2 Examination of first witness for respondent.
3 Cross examination by applicant.
4 Cross examination by members of the tribunal.
5 Steps 2, 3 and 4 are repeated for each witness.
6 Opening statement by applicant.
7 Examination of first witness for applicant.

8   Cross-examination by respondent.
9   Cross-examination by members of the tribunal.
10   Repeat steps 7, 8 and 9 for subsequent witnesses.
11   Summary of case by the applicant and then the respondent.
12   Decision announced after short adjournment. (In more complex cases the decision
     may be deferred and notified later by post).

The majority of cases heard by tribunals relate to unfair dismissal claims and in proven
cases the remedy is either reinstatement, the desirable result, or compensation which
may be set at three levels:

1   A basic award set usually at redundancy payment level but with the possibility
     of a reduction in the case of contributory responsibility.
2   A compensatory award, having a higher ceiling, takes into account re-employment
     difficulties.
3   An additional award has an element of penalty added to the compensation figure.

Appeals against the decision of a tribunal may be made to the Central Office of
Industrial Tribunals based upon the following grounds:

1   That the tribunal was in error.
2   That due notice was not served.
3   That the absence of either party caused prejudice.
4   That it would be in the interests of justice to rehear the case.

5.2.5 EMPLOYMENT APPEAL TRIBUNAL (EAT)

The Employment Appeal Tribunal was established in February 1976 under the
Employment Protection Act 1975. Its function is to hear appeals against the
judgements of tribunals and comprises a High Court Judge and a minimum of two
lay members representing commerce, industry and the trade unions. Such an appeal
must be registered within forty-two days of the judgement and in complex cases
decisions of the EAT may be taken to the Court of Appeal and then to the House of
Lords. For all these three appeals legal aid may be provided.

5.2.6 CODES OF PRACTICE

Provision for the introduction of statutory Codes of Practice was contained in the
Industrial Relations Act 1971 and is now included in both the Employment Protection
Act 1975 and the Employment Act 1980. The EPA enabled ACAS to prepare guidance
in the form of Codes of Practice for the purpose of promoting the improvement of
industrial relations, and the EA extended this right to the Secretary of State for
Employment. The Commission for Racial Equality, the Equal Opportunities
Commission and the Health and Safety Commission also may publish Codes of
Practice.

   By definition Codes of Practice are not legally binding but their strength lies in the
fact that they are admissable as evidence in any proceedings before a tribunal or the
Central Arbitration Committee, who will take into account those conditions
considered relevant.

5.2.7 THE INDUSTRIAL RELATIONS CODE OF PRACTICE 1972

This was the first Code of Practice of its kind issued under provisions contained in the
Industrial Relations Act and unless superceded by other codes or legislation it remains
valid. The Code contains six main sections, four of which remain substantively, and is
based on two main themes:

1 The vital role of collective bargaining.
2 The importance of good human relations.

Responsibilities The first section sets out the duties of management and employer associations as well as those of individual employees and trade unions.

Employment policies Practical guidance is given in the areas of

1 Planning and manpower use.	4 Payment systems.
2 Recruitment and selection.	5 Status and security
3 Training.	6 Working conditions.

Communication and consultation Certain of the references in this section have been superceded but the general guidance on the nature and importance of good communication and effective consultations are still valid.

Collective bargaining This section has been replaced by ACAS CoP 2 and additional legislation.

Employee representation at the place of work Because the recommendations are made in principal and not in detail much of this section remains valid, and is indeed built into agreed procedures in the building industry. The following aspects are covered:

1 Functions of a shop steward	4 Co-ordination
2 Appointments	5 Facilities (ACAS CoP 3)
3 Status	6 Training (ACAS CoP 3)

Grievance and disputes procedure This section has been superceded by ACAS CoP 1.

5.3 CODES OF PRACTICE ISSUED BY ACAS

5.3.1 CoP 1 DISCIPLINARY PRACTICE AND PROCEDURES IN EMPLOYMENT

CoP 1 Disciplinary Practice and Procedures in Employment came into effect on the 20th June 1977. It gives practical guidance on how disciplinary rules and procedures may be drawn up. Whilst the Code accepts that not all the provisions are practicable for a small firm, the listed essential features could be incorporated into a simplified procedure.

Why have disciplinary rules and procedures? Rules and procedures promote fairness in the treatment of individuals and in the conduct of industrial relations. They establish standards of conduct and can be used as a basis of comparison in the event of any appeal against dismissal.

Formulating the policy Whilst it remains management's responsibility to ensure the existance of rules and procedures, they will have most effect when they are established after joint consultation.

Rules It is advised that rules be clear and concise and that, whilst copies should be available for reference, they should be explained orally to all new entrants. The consequences of the breaking of rules must also be explained particularly those which might warrant summary dismissal.

Essential features of a disciplinary procedure Disciplinary procedures should:
1 Be written.
2 State to whom they apply.
3 Specify disciplinary action.
4 Indicate the authority held by varying levels of management.
5 Require notification of the complaint to the individual, allow him to state his case and allow for trade union representation if applicable.

6 Restrict dismissal for a first offence to cases of gross misconduct.

7 Provide for an effective but speedy investigation.

8 Specify the appeals procedure.

The recommended procedure to be followed in cases of a suspected breach of discipline should be:

1 The immediate establishment of the facts through the statements of witnesses and people involved.

2 In serious cases the suspension with pay of the individual.

3 The taking of a statement from the individual concerned and the notification of all rights and appeals.

The levels of disciplinary action are established as follows:

1 Informal oral warning.

2 Formal oral warning as the first stage in a disciplinary procedure, then

3 Final written warning and the advice that further breaches will lead to dismissal, then

4 Disciplinary transfer, suspension without pay if provided for, or dismissal.

5 Summary dismissal

It is recommended that special operating procedures be established for:

1 Employees who are unable to take advantage of the full procedure.

2 Trade union officials.

3 Criminal offences outside employment.

Appeals The need for speed usually requires separate appeals procedures from those adopted for the case of grievances for example. Arbitration is recommended as the final stage in any procedure.

Records A confidential but thorough record should be kept of all breaches of discipline any subsequent action.

Further action A review of all procedures and rules should be carried out periodically and amendments established after due consultation and notice.

5.3.2 CoP 2. DISCLOSURE OF INFORMATION TO TRADE UNIONS FOR COLLECTIVE BARGAINING PURPOSES

Code of Practice 2 came into effect on 22nd August 1977 and is a response to the general duty in law of an employer to disclose information requested by the trade unions to enable effective collective bargaining. To be covered by the Code information must:

1 Be in the possession of the employer and relate to his undertaking.

2 Be necessary for meaningful collective bargaining purposes.

3 Contribute to good industrial relations practice.

Its disclosure may be refused if:

1 It would be against the interest of national security.

2 It was given to the employer in confidence.

3 It relates to an individual who has refused permission for its disclosure.

4 Its disclosure would considerably damage the enterprise.

5 Its cost of production would outweigh the benefit.

If, following a written request for information, the trade union concerned consider it to be unfairly withheld then an appeal must be sent to the Central Arbitration Committee who will attempt to achieve reconciliation through ACAS. Should this fail the Committee will hear and determine the matter.

Providing information Although the Code indicates that, because of the variety of circumstances, it would be impossible to provide an exhaustive list of all required disclosures. It suggests that, if relevant, the following areas may be concerned:

1 Pay and benefits	4 Performance
2 Conditions of service	5 Finance
3 Manpower	

Restrictions to the duty to disclose Should an employer wish to resist a request for the disclosure he must establish the injury that it would disclose. The following examples are quoted of instances where a restriction may be upheld:

1 Cost information on individual products.
2 Analysis of investment, marketing or pricing policies.
3 Make up of tender prices.
4 Instances which may lead to loss of customer or suppliers or to impair the ability to raise finance.

Trade Union responsibilities Trade unions should co-ordinate their requests through trained, nominated representatives. The requests should be precise and complete, should state the reasons for the request and should allow the employer time to respond.

Employer's responsibilities Employers are recommended to respond to justified requests as quickly as possible and in a manner likely to be understood by the trade union. Any refusal must be substantiated by reasons which may well be evaluated at a future time by CAC.

Joint arrangements for the disclosure of information It is recommended that an agreed procedure for the disclosure of information is established between employer and trade union. This should preferably include an internal disputes procedure to prevent all disputes leading to adjudication by the external bodies.

5.3.3 CoP 3. TIME OFF FOR TRADE UNION DUTIES AND ACTIVITIES

This Code of Practice was enacted on the 1st of April 1978 and gives practical guidance in respect of the legal requirement for the employer to grant reasonable paid time off in order that the official might:

1 Carry out necessary relevant duties.
2 Undergo relevant approved training.

Disputes in regard to this provision are again the subject of a conciliation attempt by ACAS and if that fails a hearing before an industrial tribunal.

General considerations The Code does not lay down fixed procedures because of the variety of circumstances but suggests that agreement is reached in accordance with other procedures.

Trade Union Official's duties The following list is established as duties which might properly be carried out within working hours:

1 Collective bargaining.
2 Informing members of the results of consultations.
3 Meeting other officials on industrial relations matters.
4 Conducting interviews with members regarding grievances or disciplinary disputes.
5 Appearing at tribunals and other hearings on behalf of members.
6 Inducting new employees into the industrial relations structure.

Training The need for initial and subsequent training of trade union officials is established and it is recommended that procedures be devised for the identification of training needs and relevant courses.

Trade Union activities The right to paid time off work for the following additional activities is established:

1 Taking part in wider trade union activities.

2 Organising and participating in ballots.

3 Holding emergency meetings within working hours with management permission.

Conditions relating to time off Trade union officials must present co-ordinated requests for time off, as far in advance as possible and must bear in mind management's problems related to production, service and safety.

Industrial action Where industrial action is threatened, emergency time off may well assist in the resolution of the problem particularly where the action is involving a small number of workers.

5.4 CODES OF PRACTICE ISSUED BY THE SECRETARY OF STATE FOR EMPLOYMENT

As previously established the Employment Act 1980 empowered the Secretary of State for Employment to introduce Codes of Practice in addition to ACAS. The first two Codes came into effect on the 17th December 1980 and cover closed shop agreements and picketing.

5.4.1 CoP CLOSED SHOP AGREEMENTS AND ARRANGEMENTS

Section B. Legal rights of individuals The individual received additional protection under the Employment Act 1980. The Code sets out conditions under which it is unfair to dismiss or take action against an employee for not being a trade union member:

1 An objection on the grounds of conscience.

2 Employment in that company before the establishment of the closed shop.

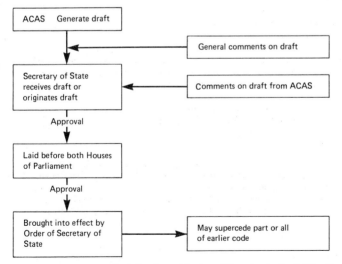

Fig 5.2 The process of origination of Codes of Practice by both The Secretary of State and ACAS

3 Where the closed shop came into being on or after 15th August 1980 not
approved by 80% of those entitled to vote.

Unfair dismissal, in these circumstances, leads to the award of compensation or
reinstatement and anyone found to have pressurised the employer is liable to become
involved in compensatory awards.

Section C. Closed shop agreements and arrangements Before negotiations for a closed
shop agreement commence, employers are advised to take account of the following:

1 The need for participation.
2 The legal implications.
3 The need for a high level of agreement.
4 The ethics and convictions of groups of workers incompatible with trade union
activity.

The Unions similarly should:

1 Consider the views of the whole of their membership.
2 Pay regard to TUC advice where more than one union is involved.
3 Recognise the existing machinery for the settlement of disputes and differences.

It is recommended that any new agreement should:

1 Indicate those concerned and the conditions under which employees may not be
required to join a union.
2 Specify a period within which membership should be taken up.
3 Investigate claims of unreasonable exclusion from a trade union before any action
is taken.
4 Specify that if an employee is expelled from a union for refusal to join industrial
action then that employee may not be fairly dismissed.
5 Set out a complaints procedure and provide for a periodic review.

A secret ballot should be held and agreement reached on the following:

1 The proposed agreement.
2 The definition of the electors.
3 The method of providing information.
4 The framing of the ballot questions.
5 The method of balloting.

The Code requires that agreements should be operated with tolerance and flexibility
and should be explained to potential employees before recruitment. Reviews of
arrangements should take place periodically or if any of the following circumstances
arise:

1 If there is clear evidence of reduced support.
2 Where a change in the parties to the agreement occurs.
3 If the arrangement does not work satisfactorily.
4 Where there is a change in the law.

Section D. Union treatment of members and applicants A requirement is imposed on
the unions to establish fair and clear rules and procedures relating to membership and
discipline. Disciplinary action should not be taken against members who, in certain
circumstances, refuse to take part in industrial action or who cross picket lines.

The last section of this code deals specifically with the problems of journalists and the
freedom of the press.

5.4.2 CoP PICKETING

Section B. Picketing and the Civil Law. Picketing is only lawful if:

1 It is concerned with a trade dispute.

86

2 It is carried out adjacent to a person's place of work, except for accredited trade union officials in certain circumstances.
3 It involves peaceful persuasion or the transmission of information.
4 It is lawful secondary action where employees are picketing their employer in support of a dispute between another employer and his employees.

The remedies for illegal picketing allow the employer to take an action for damages and request an order to prevent further picketing.

Section C. Picketing and the criminal law The immunities described do not extend to the criminal law and a picket is liable to arrest if he:
1 Intimidates by language.
2 Intimidates by behaviour.
3 Possesses an offensive weapon.
4 Damages property.

Pickets have no right to stop vehicles but may request that they do by words or gesture, and the driver of a vehicle approaching a picket line may not drive in a manner likely to inflict injury.

Section D. The role of the police The police will:
1 Uphold the criminal law.
2 Restrict the number of pickets.
3 Use discretionary powers to prevent a breach of the peace.
4 Assist, if they consider that there might be a breach of the peace, officials who are in the process of serving court orders resulting from civil action.

Section E. Limiting the numbers of pickets Excessive numbers of pickets are considered by the Code to be the major cause of violence and disorder. It is recommended that no more than six pickets are present at any one entrance and suggested that on occasions a smaller number will suffice.

Section F. The organisation of picketing The picket should preferably be organised by a trade union official with written authorisation. His job is to:
1 Ensure an understanding of the law.
2 Distribute badges and armbands.
3 Refuse offers of assistance from anyone except those who work at the picketed premises.
4 Liaise with his own and other union officials.
5 Ensure the maintenance of essential supplies.

Section G. Essential supplies and services Distress, hardship and inconvenience should not be caused to people not involved in the dispute. The Code suggests the following illustrative list of essential supplies:
1 Medical and pharmaceutical products.
2 Supplies to health and welfare institutions.
3 Heating fuel for schools and houses.
4 Essential services, police, fire, ambulance etc.
5 Livestock.

5.4.3 CoP. THE ELIMINATION OF RACIAL DISCRIMINATION AND THE PROMOTION OF EQUALITY OF OPPORTUNITY IN EMPLOYMENT

It is expected that this Code of Practice, prepared for the Secretary of State for Employment by the Commission for Racial Equality, will be put to the House of

Commons in 1982. The draft sets out measures which must be taken by employers, trade unions and employment agencies to ensure a greater equality of opportunity. The Code covers the following:

1 General policy.
2 Recruitment criteria.
3 Interviewing.
4 Cultural needs.
5 Training.
6 Monitoring of the ethnic composition of the workforce.
7 Undertaking of positive action where there is under-representation of particular racial groups.

BIBLIOGRAPHY

ACAS: (All the following are published by HMSO)
Code of Practice 1: *Disciplinary Practice and Procedures in employment.*
Code of Practice 2: *Disclosure of information to Trade Unions for collective bargaining purposes.*
Code of Practice 3: *Time off for Trade Union duties and activities.*
The ACAS Role, (1978)
Industrial relations handbook (1980)
Croner's Reference Book for Employers, Amendment Service, Croner Publications Ltd.
Department of Employment: (The following are published by HMSO)
Code of Practice: *Closed shop agreements and arrangements*
Code of Practice: *Picketing*
Industrial relations Code of Practice, HMSO.

EXERCISES

RECALL QUESTIONS

1 Complete the following Acts relevant to industrial relations
  (a) .... Protection Act 1975
  (b) Trade Union and L ..... R .... Act 19 ..
  (c) .... Act 1980
  (d) .... (Consolidation) Act 1978
  (e) E .... Act 1970

2 Identify the following bodies concerned with industrial relations
  (a) CO   (d) I.T.
  (b) ACAS   (e) EAT
  (c) CAC

3 List the stages in the origination of Codes of Practice by ACAS
  (a) D .... the Code of Practice
  (b) Submit to the S .... of S .... for E ....
  (c) Lay before both H ..... of P ....
  (d) Publish the Code of Practice

4   List the recommended stages of progressive disciplinary action
    (a);
    (b);
    (c);
    (d).

5   Identify the three levels of financial award granted by Industrial Tribunals in cases
    of proved unfair dismissal
    (a) A . . . . award; (b) B . . . . award; (c) C . . . . award

QUESTIONS REQUIRING SHORT ESSAY ANSWERS (15—20 minutes)

1   Describe a likely procedure to be followed by an Industrial Tribunal in cases of
    unfair dismissal.

2   Explain in detail the two methods by which Codes of Practice may be originated.

3   Identify the situations within industrial relations which might result in the
    involvement of ACAS.

4   As a trade union official prepare, and justify, a request for the disclosure of the
    company's 5-year manpower plan.

5   Indicate those duties and activities which may, by agreement, be conducted by a
    trade union official in working hours.

QUESTIONS REQUIRING LONGER ESSAY ANSWERS (30—45 minutes)

1   Trace the development of industrial relations legislation between 1970 and 1980 in
    relation to the changes in government.

2   Prepare a report for your Company recommending a disciplinary procedure which
    meets the requirements of CoP 1 Disciplinary Practice and Procedures in Employment.

3   Define 'Closed Shop Agreement' and consider the benefits and drawbacks to
    employer and employee.

4   Differentiate Laws and Codes of Practice as means of achieving good industrial
    relations.

5   As a convenor steward you intend to organise a picket of the site upon which you
    are employed during a national official strike.
    Detail the arrangements you would make to ensure that the picket complies with
    the requirements of the Code of Practice on Picketing.

6 The National Working Rules

6.1 The National Working Rules	

| 6.2 Health and Welfare Conditions for the building industry | 6.2.1 Provision of First Aid boxes or cases
6.2.2 Standard of training in First Aid treatment
6.2.3 Ambulances
6.2.4 First Aid rooms
6.2.5 Application to office workers
6.2.6 Shelter and accommodation for clothing and taking meals
6.2.7 Washing facilities
6.2.8 Sanitary conveniences
6.2.9 Protective clothing
6.2.10 Safe Access to facilities |

| 6.3 Agreement for annual holidays with pay scheme | 6.3.1 Management of the scheme
6.3.2 Operation of the scheme |

| 6.4 Benefit scheme for the building and civil engineering industry | 6.4.1 The Death Benefit Scheme
6.4.2 The Retirement Benefit Scheme |

| 6.5 Other sections | |

The National Working Rules for the Building Industry (NWR), are published by the NJCBI (as established in chapter 4) in a booklet referred to as the *Working Rule Agreement*. They constitute an agreement between employers affiliated to the NFBTE and the trade unions FTAT, GMWU, TGWU and UCATT. The rules establish the minimum terms and conditions of employment within the building industry but for historical and regional requirements it is necessary to publish four editions:

1 The General Edition.   3 The Liverpool and District Edition.
2 The Scottish Edition.   4 The London Edition.

It becomes necessary also to amend the rules from time to time, and it is intended to issue the document annually following each wage settlement.

This chapter is based on edition A, the General Edition which became effective on 4th October, 1982. However, even within this edition there are Regional Variations and additional working rules included in a separate section.

It is important therefore to recognise that, whilst this chaper deals with principle rather than detail, the reader must make reference to the actual document and ensure that:

1 It is the edition relevent to the area of operations,
2 The latest amendments are included.

6.1 THE NATIONAL WORKING RULES

There are twenty seven rules included in the document each of which is summarised below. Some of the rules are followed by guidance notes and many are amended by Regional Variations included in a separate section.

NWR 1 Guaranteed minimum weekly earnings Definitions are set out regarding the entitlement of building trade operatives to minimum weekly earnings which include a guaranteed minimum bonus payment.

NWR 2 Basic rates Basic rates are set out for craft operatives and adult male labourers with variations for

Trade chargehands and gangers;
Qualified first-aiders;
Qualified benders and fixers;
Maintenance fitters;
Watchmen;
Young labourers;
Workers in woodworking establishments.

NWR 3 Extra payments for continuous extra skill and responsibility
Payments above the labourers basic rate are specified for those with responsibility for the operation of plant, pipelayers and jointers and concrete workers.

NWR 4 Extra payments for intermittent responsibility Additional payments are specified for labourers involved in work on simple access scaffolding and on certain stone cleaning operations.

NWR 5 Payment arrangements The basis of payment, pay week and pay day are specified.

NWR 6 Working hours The timing and duration of the normal working day and meal intervals are indicated.

NWR 7 Overtime General provisions are made in respect of overtime working. The method of calculating additional payments in respect of the normal working day is established.

NWR 8 Shiftwork Double day shift and three shift working is defined and the basis for the calculation of additional payments specified.

NWR 9 Nightwork Nightwork is differentiated from shiftwork and additional payment set out for nightworkers.

NWR 10 Annual holidays The duration and means of fixing the dates of the annual holiday entitlement are set out.

NWR 11 Public Holidays The entitlement to recognised public holidays with appropriate payment is established.

NWR 12 Transfer arrangements This rule provides the employer with the authority to transfer an operative from one job to another.

NWR 13 Termination of employment The rights of notice and conditions and procedures for dismissal are established. These are detailed in section 2 of Chapter 7.

NWR 14 Daily travelling Rules and payments for travelling to sites on a daily basis are specified. The calculations and payments involved in this section are complex and separate explanatory notes are included to detail the rights to, and size of payments in respect of both expenses and travelling time.

NWR 15 Periodic leave and lodging Entitlements in respect of work on distant sites to which the operative does not travel on a daily basis are specified.

NWR 16 Sickness and injury payments Provided that certain conditions have been met, a daily payment is specified for absences in respect of certificated illness and injury.

NWR 17 Payment for work in difficult conditions Additional hourly payments are listed for work in conditions of discomfort, inconvenience and risk.

NWR 18 Tool and clothing allowances Additional weekly payments are specified for the provision and maintenance of a list of mandatory tools.

NWR 19 Benefits scheme The Building and Civil Engineering Benefit Scheme, details of which are included in section 4, establishes an entitlement to payments in respect of death or retirement.

NWR 20 Conditions of training of apprentices trainees The requirements for trainees and employers to enter into a Training Service Agreement is specified.

NWR 21 Conditions of employment of apprentices/trainees The scale of payments made to apprentices and trainees under all headings is specified.

NWR 22 Scaffolders This rule differentiates between Trainee, Basic and Advanced Scaffolders and establishes levels of additional payments.

NWR 23 Wearing of Safety helmets In furtherance of the objectives of The Health and Safety at Work etc. Act, employers are required, after due consultation, to reach agreements in respect of the wearing of safety helmets. Where such agreements exist the specified disciplinary procedures must be supported by both unions and employers.

NWR 24 Safety representatives The appointment, training and functions of Safety Representatives are covered. Details of this rule are outlined in section 2 of Chapter 8.

NWR 25 Trade union recognition and procedures Much of the detail included in this rule is to be found in section 5 of Chapter 4.

NWR 26 Register of employers Employers registered with the NJCBI undertake to
1 Establish direct communications with unions on appropriate matters.
2 Attempt to ensure that all operatives are directly employed (as opposed to labour only subcontracting).
3 Notify union officials of the commencement of new contracts and the likely demand for operatives.
4 Be prepared to deduct union subscriptions at source.

NWR 27 Grievance, disputes and differences This rule is covered in detail in section 7 of Chapter 4.

6.2 HEALTH AND WELFARE CONDITIONS FOR THE BUILDING INDUSTRY

The Construction (Health and Welfare) Regulations 1966 as amended by the Construction (Health and Welfare) (amendment) Regulations 1974 require the provision of a minimum level of facilities dependent generally upon the number of operatives employed on site.

As there is generally more than one employer or contractor on a site the regulations allow for the duty of provision to be shared. In such cases the number of operatives is to be considered as to total number using that facility and a register must be maintained listing the sharing firms.

On July 1st 1982 the Health and Safety (First Aid) Regulations 1981 became operative and all current regulation is interpreted within the WRA as follows.

6.2.1 PROVISION OF FIRST AID BOXES OR CASES

First aid boxes, identified now with a white cross on a green background, must be provided for access by all workers. The contents are specified in varying quantities dependent upon the number of employees serviced.

6.2.2 STANDARD OF TRAINING IN FIRST AID TREATMENT

Number of operatives	First aid personnel
up to 50	An 'appointed person' to handle emergencies, summon help and take charge of equipment.
50–150	At least one first aider.
151 +	An additional one first aider for every 150 employees.

A first aider holds a first aid certificate less than three years old from a specified authority or has been trained to a level acceptable to the Health and Safety Executive.

Sites with special risks and those spread over a wide area require special treatment.

6.2.3 AMBULANCES

If more than twenty-five operatives are employed the contractor must notify the ambulance authority of the site address, the nature of the work and the projected completion date. At least one stretcher must be provided, and an individual must be identified and trained to make emergency telephone calls for an ambulance. This person is to be available at all working hours and his name must be displayed on notices. If a telephone or radio link is impracticable then a vehicle capable of carrying a stretcher, and containing the name and address of the nearest emergency hospital, is to be provided.

6.2.4 FIRST AID ROOMS

A first aid room, adequately constructed, conveniently sited and specifically equipped must be provided if 250 or more operatives are employed. The room is to be in the charge of a qualified first aider.

6.2.5 APPLICATION TO OFFICE WORKERS

Site office staff must have access to all first aid facilities and in the calculation for operative usage three office workers count as one operative, any remaining one to be counted down and two to be rounded up.

6.2.6 SHELTERS AND ACCOMMODATION FOR CLOTHING AND FOR TAKING MEALS

Number of operatives	Required Provision
All sites	Shelter to be provided, with such warming and drying facilities as are practical for:
	Bad weather protection
	Personal clothing
	Protective clothing
	Meals (seats and hot water)
	All shelters are to be kept clean and drinking water is to be provided and so marked.
6+	Shelters to be warmed and have drying facilities.
11+	A means of heating food must be provided if hot food is not available on site.

All shelters must be conveniently sited but available transport can be taken into consideration. In any calculations the number of operatives is taken to be the number likely to use the facility at any one time.

Required provision

All sites longer than 4 hours duration.	Washing facilities are to be provided.
More than 6 weeks *or* 21+ operatives	Hot and cold (or warm) water to be provided in receptacles along with soap and towel or driers.
More than 12 months *and* 101+ operatives	All as above but with a specified number of receptacles

On any site where lead or other poisonous compounds are used, nail brushes must be provided along with the facilities listed immediately above.

6.2.8 SANITARY CONVENIENCES

On all sites one convenience must be provided for every twenty-five operatives. If however 101 or more operatives are employed and screened urinals are provided then four conveniences plus one for every additional thirty-five men are required. Facilities must be separated for men and women and they must be accessible, partitioned and have fastening doors. They must be adequately lit and ventilated and be kept clean and must not open directly into a mess or work room.

6.2.9 PROTECTIVE CLOTHING

Operatives who have to work in rain, sleet snow or hail must be provided with protective clothing.

6.2.10 SAFE ACCESS TO FACILITIES

All facilities must be maintained in good order and be reached by proper paths.

6.3 AGREEMENT FOR ANNUAL HOLIDAYS-WITH-PAY SCHEME

This agreement originally established in October 1942 but subjected to revision, seeks to provide holidays with pay for those employed in the building and civil engineering industries.

6.3.1 MANAGEMENT OF THE SCHEME

The scheme is operated by the Building and Civil Engineering Holidays Scheme Management Ltd and applies to all operatives covered by the WRA of both NJCBI and CECCB. Apprentices are excluded from the scheme except that in their final year provision must be made in order that they can receive payment in their first year as craft operatives.

The scheme is operated by the attachment of stamps to a card which is personal to each operative. Whilst he remains in employment his card is held by his employer who attaches additional stamps for each week of approved employment. At the commencement of the holiday period in the succeeding year the stamps are cashed by the employer and the money paid to the operative.

The management of the scheme reimburses the employer within 14 days of receipt of the official payment declaration form and the relevant stamping cards. If an operative is incapacitated, unemployed or working outside the industry he may claim payment directly at the appropriate time. The responsibility for interpretation of the scheme rests with the BCEJB who may vary or terminate the scheme.

6.3.2 OPERATION OF THE SCHEME

On the first Monday of April in each year an employer obtains a numbered holiday card for each operative in his employment. During the year stamps are purchased by the employer and fixed in the spaces provided. Subject to the provisions of the relevant WRA stamps are provided for each operative working a full or part week, or who, after a minimum of 5 days paid employment, is on certificated sick leave whether or not he is receiving sick pay. For each of these weeks the stamp is cancelled with a letter S. Pay is provided for:

1 A Winter Holiday of seven working days which in connection with Christmas Day, Boxing Day and New Years Day provide two weeks.
2 An Easter (Spring) Holiday of four working days following Easter Monday (or Spring Holiday Monday in Scotland) to give one week.
3 A Summer Holiday of two weeks between 1st April and 30th September which need not be consecutive.

The timing of the summer holiday may be determined by custom or by request, both subject to negotiation should there be any overriding difficulties. During the last pay-week prior to the commencement of the holiday the employer will make a cash payment to the value of the stamps on the appropriate section of the card. Should the number of stamps fall below a stated minimum then a payment may be made in respect of stamps provided for a future holiday period.

6.4 BENEFIT SCHEME FOR THE BUILDING AND CIVIL ENGINEERING INDUSTRY

6.4.1 THE DEATH BENEFIT SCHEME

This agreement, initiated in November 1974, seeks to provide a death benefit for the spouse or dependents of a construction operative who dies from any cause whilst employed under the NJCBI or CECCB Agreements.

The scheme is operated through a Fund under a Trust Deed and legally administered by the Building and Civil Engineering Holidays Scheme Management Ltd. It applies to all operatives between the ages of 18 and 65 who are covered by the WRA of the NJCBI or CECCB but certain elements are extended to apprentices below the minimum age limit. Payments are made into the scheme by way of an addition to the value of stamps purchased under the holidays with pay scheme.

Following the death of an operative in service, a claim must be initiated by his spouse or dependents which should be forwarded in the appropriate manner to the Trustees within three months of the death.

An operative is covered for benefit, subject to the age limits mentioned, if he is actively employed under the WRA and has a minimum of four stamps attached to his holiday card in the eight weeks prior to his death.

In the case of absence for holidays, lay-off or sickness this period may be excluded from the 8 weeks. If an operative is covered on his last working day before aholiday or a temporary lay-off then payment will be made in respect of death during that period.

He will also be covered for a period of four weeks after the termination of employment provided that he was covered on his last working day.

6.4.2 THE RETIREMENT BENEFIT SCHEME

This scheme was introduced from April 5th 1982 to provide a lump sum retirement benefit for operatives who have reached the age of 65 or who are taking specified early retirement. As an interim measure an operative with eighty stamps or more on his holiday with pay card in the previous two years is entitled to a cash benefit of up to £564. When the scheme becomes fully operational the sum will be dependent upon 'reckonable service'.

6.5 OTHER SECTIONS

In addition to the detail described the WRA contains other sections which deal with:
1 The industrialisation of the building process.
2 General principles concerned with incentive schemes and productivity agreements.
3 Racial discrimination in employment.
4 The areas covered by Regional Joint Committees.
5 The National Joint Training Scheme for skilled building occupations.

BIBLIOGRAPHY

National Joint Council for the Building Industry, *National Working Rule Agreement (1982 edition) NJCBI, London.*

EXERCISES

Are you in possession of an up to date Working Rule Agreement relevant to your area of operations? Use it to answer the following:

RECALL QUESTIONS

1 Complete the headings of the rules with the following numbers
   (a) NWR 2 . . . .
   (b) NWR 9 . . . .
   (c) NWR 17 . . . .
   (d) NWR 22 . . . .
   (e) NWR 27 . . . .

2 List all the possible components of a craft operatives pay.

3 List the facilities required by health and welfare regulations for a site employing fifty-one operatives.

4 A site has approval to work a nine-hour day and four hours on a Saturday morning.

Calculate the number of hours, including the overtime premium, which the operative will be paid.

5　　An operative, who is entitled to allowances when travelling to sites on a daily basis travels an allowable 17 km on Monday and 11 km on the other 4 days. Calculate his total allowances.

QUESTIONS REQUIRING SHORT ESSAY ANSWERS (15–20 minutes)

1　　Outline the procedure whereby building operatives receive holiday payment.

2　　List the variations and additional working rules applicable to your site or the area of operation of your company.

3　　Differentiate between the qualifications of, and work carried out by: Trainee Scaffolders; Basic Scaffolders and Advanced Scaffolders.

4　　Compare modern temporary accommodation with traditional sectionalised timber hutment as means of providing health and welfare facilities on building sites.

5　　Differentiate between double day shift, three shift and nightwork indicating the type of contract which might require each work pattern.

QUESTIONS REQUIRING LONGER ESSAY ANSWERS (30–45 minutes)

1　　When set, minimum standards tend to become the maximum provision. Discuss this statement in respect of health and welfare conditions in the construction industry.

2　　In addition to his basic rate of pay an operative may receive many extra payments. Analyse in detail all such payments.

3　　Examine in detail the Health and Safety (First Aid) Regulations 1981 and identify the changes which became necessary to the WRA.

4　　The cost of health and welfare facilities must be included in the competitive element of a contractor's tender. Analyse the dissadvantages of such a system and suggest a superior alternative.

5　　List the conditions under which an operative becomes eligible for allowances in connection with his travel to work. Detail the method of calculating such allowances.

7 The engagement and termination of employment, disputes and differences

7.1 Engagement	7.1.1 Written particulars of employment 7.1.2 Itemized pay statement

7.2 Termination of employment	7.2.1 Site transfer 7.2.2 Dismissal procedures 7.2.3 Summary dismissal 7.2.4 Periods of notice 7.2.5 Unfair dismissal 7.2.6 Remedies for unfair dismissal

7.3 Redundancy	7.3.1 Temporary lay-off 7.3.2 Calculation of redundancy payment 7.3.3 The redundancy fund 7.3.4 Redundancy procedures

7.4 NWR 27 Grievances, Disputes & Differences	7.4.1 Grievance procedures 7.4.2 Disputes and differences 7.4.3 Rule 8 NJCBI Rules and Regulations

The Employment Protection (Consolidation) Act 1978 as amended by the Employment Act 1980 established certain rights for the individual (see chapter 5). These rights become the minimum legal requirements and cannot therefore be eroded by any in-company or national agreement. There is however no reason why such agreements may not afford improved terms and conditions of employment to employees. The statutory rights of individuals are considered in three sections:

1 Engagement.
2 Termination.
3 Redundancy.

Successive codes of practice have stressed the requirement for formal disputes procedures and the final section of this chapter deals with the requirements of the WRA in respect of such procedures.

7.1 ENGAGEMENT

7.1.1 WRITTEN PARTICULARS OF THE TERMS OF EMPLOYMENT

The induction procedures outlined in Chapter 3 indicated that certain documentation would be exchanged. Provided that they are to work a minimum of 16 hours a week, employees must be given either a written statement or a written contract of employment specifying the particulars listed below. This information must be given no later than 13 weeks after the date of engagement which must be specified on the document in addition to the names of both employee and employer and the date of expiry if it is for a fixed term. Other information which must be included is:

1 The rate of pay, method of calculation and whether it is hourly, weekly or monthly.
2 Normal working hours and relevant rules.
3 Holiday entitlement and payment.
4 Sickness and absence rules and payment.
5 Details of pension scheme and relationship with state pension.
6 Required notice on the part of both parties.
7 The job title.

Details must also be included of:

1 Disciplinary rules and grievance procedures.
2 Persons to whom grievances can be addressed and appeals made.

If much of the detail is included in reference documents then it does not have to be individually noted. It is sufficient to give the name of the document to every employee and inform them where it can be inspected. Changes must be confirmed within 1 month of them becoming operative by a note or by a revised document.

If an employee has not received a statement or has other specified complaints he may apply to a tribunal, which may redress the problem but is unable to impose financial penalties.

Organisations may produce or purchase standard forms for this purpose, an example of which is given in *Fig 7.1*.

7.1.2 ITEMIZED PAY STATEMENT

Employees must receive on or before the pay date a statement showing the following:

1 Gross amount of wage or salary.
2 Nett amount of wage or salary.

TERMS OF EMPLOYMENT STATEMENT

Employer's Name

Address

Employee's Name

Address

Job Title   Commencement Date

Remuneration

Working Hours

Holidays and Holidays with Pay

Pension Rights

Sickness and Injury

Notice of Termination

Disciplinary Rules and Procedures

Grievance Procedures

Signed ....   Issued on ....

for and on behalf of the Employer

Please also see attached 'Notes for Guidance'

Fig 7.1 Terms of Employment statement

3 Reasons for and amount of variable reductions.
4 Reasons for and amount of fixed reductions.
5 Explanation of the make up of the gross amount if in different parts.

A tribunal may compensate an individual for unauthorised deductions backdated to 13 weeks prior to the application.

7.2 TERMINATION OF EMPLOYMENT

The temporary nature of work on individual construction sites has traditionally created an environment of casual labour within which site management was able to 'hire and fire' at will. Gradually the WRA developed protection for the employees of the industry who were subject to its conditions and eventually the national legal framework extended similar protections to all. The detail of this section relates to NWR 12 Transfer Arrangements and NWR 13 Termination of Employment which affords better terms and conditions generally than the statutory minimum and also contains rules for circumstances specific to the building industry.

7.2.1 SITE TRANSFER

Any operative can, at the discretion of the management, be transferred from one site to another under the financial provisions of NWR 14 – Daily Travelling. Should he be required to lodge then his consent is necessary unless, extenuating circumstances apart, he has lodged away from home within the preceding 12 months with the same employer.

7.2.2 DISMISSAL PROCEDURES

In accordance with the Code of Practice, the procedure for dismissal in cases other than gross misconduct or where the individual concerned has been employed for less than 6 normal working days, is as follows:

1 An oral warning should be issued if an operative fails to meet standards of conduct and workmanship required. It is preferable for this warning to be given before a witness.
2 Any second warning found necessary should be given in writing. If a shop steward is involved the union should be notified.
3 If the individual repeats the offence which resulted in the warnings he should then be dismissed.

In all cases the reasons for the dismissal must be clearly stated in writing.

7.2.3 SUMMARY DISMISSAL

The WRA provides for the summary discharge of any operative at any time for 'misconduct' but the law makes it clear that summary dismissal should be reserved for only the gravest of instances in which no reasonable employer could tolerate the individual's continued employment.

7.2.4 PERIODS OF NOTICE

The notice required to be given by each side for those covered by WRA and national legislation only is listed in *Table 7.1*.

Table 7.1 Periods of notice of termination of employment

Length of continuous employment	NWRA National Working Rule 13.1		Employment Protection (Con.) Act 1978 as amended by The Employment Act 1980	
	Employer	Employee	Employer	Employee
During the first 6 days	2 hours to expire at the end of the normal day	Same	Nil	Same
Over 6 days – 4 weeks	1 day to expire end of normal Friday	Same	Nil	Same
Over 4 weeks – 2 years	1 week	Same	1 week	Same
Over 2 years – 12 years	1 week for each full year	1 week	1 week for each full year	1 week
Over 12 years	12 weeks	1 week	12 weeks	1 week

7.2.5 UNFAIR DISMISSAL

The law protects most classes of employees against unfair dismissal providing they have worked for 16 hours a week or more for a minimum of 52 weeks (or between 8 and 16 hours a week for 5 years). For a company employing less than 20 people the qualifying period is 2 years provided that the employee concerned was employed on or after October 1 1980. Dismissal is defined as:

1 The termination of a contract by the employer, with or without notice.
2 The termination of a fixed term contract before the expiry of the term.
3 The termination of a contract by the employee because of aggravated behaviour by the employer (Reduced pay, increased hours, changing the nature of work, all contrary to agreement and without consent.) This condition is called 'forced resignation' or 'constructive dismissal'

An employee who considers that he has been unfairly dismissed may lay the complaint before a tribunal using form IT 1. The form must be submitted, where reasonable, no later than 3 months after the dismissal whereupon an ACAS Conciliation Officer will endeavour to negotiate a satisfactory outcome.

The employee must prove that he was dismissed but once this is established it is up to the employer to prove that the dismissal was fair in relation to one of the following reasons:

1 The capability or qualification for carrying out the work.
2 The employee's conduct.
3 Redundancy.
4 A legal restriction making continued employment impossible.
5 Other substantial reason.

At this point the issue is generally one of 'does the punishment fit the crime?'

and 'was the manner of the dismissal fair?' The ACAS Code of Practice on Disciplinary Practice and Procedures (Chapter 5) would be considered by the tribunal at this time. There can be no fixed rules by which the tribunal is guided as each case is dealt with on its merits, but the following points of guidance can be abstracted from past decisions:

1 Employees should rarely be dismissed for their first offence.
2 A full enquiry must be carried out before dismissal.
3 The case will be decided upon the facts available at the time.
4 A 'reasonable belief' may be justification for dismissal.
5 The decision is based upon the actions of a 'reasonable management'.
6 The employee must know that he may be dismissed for the committed offence.

There are certain cases where the law is particularly complex or specifies additional protection. In the following conditions it is recommended that advice be taken.

1 Reasons connected with trade union activity.
2 Pregnant employees.
3 Dismissal of temporary replacement staff.
4 Dismissal connected with industrial action.
5 Dismissal following pressure from any party.

7.2.6 REMEDIES FOR UNFAIR DISMISSAL

If required by an employee, the first aim of the tribunal, having reached a decision of unfair dismissal, is to seek reinstatement or re-engagement. Only if this is impracticable should compensation be awarded. If the employer does not comply with a resinstatement order then an additional award may be made against the employer.
Compensation has two elements

1 A basic award calculated as redundancy pay (section 3 of this chapter) Maximum at March 1982 £3900.
2 A compensatory award covering actual financial loss (£6250)

The combined award may be reduced by the tribunal if the employee has refused reinstatement or if his behaviour is considered to have contributed to the dismissal.

7.3 REDUNDANCY

Redundancy has been established as a fair reason for dismissal. It is defined as pertaining to a case where:

1 The employer has ceased or intends to cease to carry on the business for which the employee was engaged.
2 The employer has ceased or intends to cease to carry out the business in the place where the employee was employed.
3 The work for which the employee was engaged has or is expected to cease or diminish.

It is evident therefore that construction workers will inevitably be in a position of redundancy when the demand for their skill on any contract has diminished or ceased. It is possible however for redundancy to lead to unfair dismissal if:

1 The employees are unfairly selected for redundancy.
2 The manner of implementing the redundancy is unfair.

An employer may avoid redundancy by offering suitable alternative employment and this may be done by means of a site transfer. What a 'suitable' alternative is will in these circumstances depend upon the lodging or travelling requirement.

The necessary dependence of construction work upon conditions sometimes leads to periods during winter when work is not available. The provisions to cover this in relation to redundancy is covered in both NWR 1.4 and 13.3.3 and the law, and are summarised in *Table 7.2*. If, in any period of thirteen weeks an employee has been laid off or worked on short time for six weeks of which three were continuous, then a redundancy claim may not be refuted.

Table 7.2 Timetable related to temporary lay-off

Pay week	action
1	Work ceases
2	If operative available for work, paid guaranteed minimum weekly earnings.
3	Operative may be required to register as unemployed.
4	Operative may state notice of intention to claim redundancy after 4 weeks lay off
7	Operative claims redundancy. Employer may refute claim if work is expected to resume within 4 weeks and expects the work to last for 13 weeks.

7.3.2 CALCULATION OF REDUNDANCY PAYMENT

With the exclusion of certain limited groups, employees with two years continuous service after the age of 18, when made redundant, are entitled to a tax free payment calculated by reference to their service and pay at the point of redundancy. *Table 7.3* summarises the scale.

Table 7.3 Scale of calculation of redundancy payment

Age (inclusive)	No. of weeks pay for each full year of employment
18–21	½
22–40	1
41–64 (Male) 59 (Female)	1½
	Maximum reckonable pay £130
	Maximum reckonable service 20 years
	Maximum payment £130 × 20 × 1½ = £3900 (as at March 1982)

If an employee is made redundant after the age of 65 (60 for a woman) no payment is made and if redundancy occurs in the previous year a reduction of 1/12 of the payment will be made for each completed month of service after the 64th birthday (59 for a woman).

A week's pay is considered to be the payment for a normal week's work, but if a payment by results scheme leads to wide fluctuations an average of the last twelve weeks may be taken.

7.3.3 THE REDUNDANCY FUND

Under the Redundancy Payment Rebates Regulations 1965, an employer may claim a 41% rebate of any statutory redundancy payment made. The claim is submitted on forms RP1 and RP2 after the following periods of notice; 21 days before the first redundancy if 10 or more employees are to be made redundant within a 6 day period, or 14 days in other cases. The claim must be made within 6 months of the payment being made and must be accompanied by form RP3 which is a receipt signed by the employee concerned.

The redundancy fund is maintained by a surcharge on the national insurance contribution made by employers and in effect therefore each employer is reimbursed from his own contribution. An industry which incurs considerable redundancy is subsidised by those who do not. It is for this reason that some employers in the construction industry, where payments are low because of the short periods of continuous employment, consider the scheme to be unfair.

7.3.4 REDUNDANCY PROCEDURES

If an employer proposes to make employees redundant he must notify the Department of Employment and the trade union which represents those employees. The time scale of the notice is set out in *Table 7.4*.

Table 7.4 Notice of redundancy required by the Department of Employment

No. of operatives to be made redundant at one establishment	Over a period of (days)	Minimum notice (days)
100 or more	90	90
10–99	30	60
Less than 10	As soon as possible	

In order that the trade union may meaningfully contribute to the situation additional information is required:
1 Reason for redundancy.
2 Numbers and types of employees.
3 Total number of similar employees at the same establishment.
4 Total number of all employees.
5 Proposed method of selection.
6 Proposed method of carrying out the dismissals.
Failure to undertake these statutory notifications can result in an additional payment to the employee, a reduction in the redundancy rebate or in a fine.

7.4 NATIONAL WORKING RULE 27 — GRIEVANCES, DISPUTES AND DIFFERENCES

7.4.1 GRIEVANCE PROCEDURE

Initially an individual complaint or issue should be taken up with the persons immediate superior. If the issue cannot be settled then the union steward may be brought in and the matter raised with the manager, agent/general foreman or any other person designated by management to receive complaints.

If the matter affects a group of workers, the union steward (or convenor steward if more than one union is involved) should take the matter up with the manager or appropriate person directly.

7.4.2 DISPUTES OR DIFFERENCES

If the grievance is between parties affiliated to the NJCBI and the previous procedure has not produced a solution then the case should be reported to the full-time trade union official or to the Operatives Regional Joint Secretary. Failure to agree at this level will result in the instigation of procedures set out in Rule 8 of the Constitution and Rules of the NJCBI (*not* NWR 8).

It is agreed by both sides that until the formal disputes procedure is exhausted there shall be no stoppage of work, restriction of hours worked or reduction of output.

7.4.3 RULE 8, NJCBI RULES OF THE COUNCIL

If a dispute remains unsettled after the procedure set out has been exhausted then the matter must be referred to the Local Joint Committee who are required to meet and reach agreement on a settlement within 21 days. An appeal may be made by either side within 7 days of the decision addressed to the Regional Joint Secretary.

Within 21 days the Regional Joint Conciliation Panel will meet and their decision again may be the subject of an appeal which this time should be made within 21 days to the Clerk to the National Joint Council. On receipt of this appeal, though this time with no time limit, the National Conciliation Panel will hear the dispute and give a final adjudication.

BIBLIOGRAPHY

Croner's Reference Book for Employers, Amendment service, Croner Publications Ltd
National Joint Council for the Building Industry. *Working Rule Agreement* (1982 edition) NJCBI, London.

EXERCISES

RECALL QUESTIONS

1   List 5 particulars which must appear on a written statement of employment.

2   Indicate the period of notice required to be given, under the WRA, to employees having completed the following periods of continuous employment:

| (a) 4 days | . . . . | (c) 10 years | . . . . |
| (b) 1 year | . . . . | (d) 15 years | . . . . |

3 Identify four bases of legal dismissal:

| (a) C . . . . | (c) R . . . . |
| (b) C . . . . . | (d) L . . . . R . . . . |

4 Calculate the reckonable service of an employee who has been made redundant if he has fifteen full years of actual service between the ages of:

(a) 16–30	. . . . .
(b) 23–37	. . . . .
(c) 42–56	. . . . .

5 List the information which must be provided on a pay statement.

QUESTIONS REQUIRING SHORT ESSAY ANSWERS (15–20 minutes)

1 Examine the normal and informal grievance procedures operating within your company for technical staff.

2 Explain the term 'constructive dismissal' giving realistic examples from the building industry.

3 Examine in detail the remedies for unfair dismissal and explain how the amount of any financial award will be calculated.

4 An operative is claiming redundancy following temporary lay-off. Prepare two possible calendars of events which would lead to the request being granted in one case and rejected in the other.

5 Analyse a pay statement prepared by your company for a craft operative indicating how it meets the statutory requirements.

QUESTIONS REQUIRING LONGER ESSAY ANSWERS (30–45 minutes)

1 The site upon which you are working is approaching completion and of the twelve operatives remaining, two are craft apprentices, six have been continuously employed for periods of between three and eight years and four were engaged on that contract which lasted for sixteen months.

Discuss the position and procedures in relation to the possible termination of employment of each category of employee.

2 'The Redundancy Payments Scheme is financially unfair to the construction industry'. Discuss.

3 Explain in detail the disputes procedure pertaining to operatives working under the NJCBI from its origins on site to a hearing by the National Conciliation Panel.

4 As a personnel officer you are worried about the way in which the general foremen carry out their responsibilities in relation to dismissals for reasons other than redundancy. Prepare a series of guidance notes which will assist them to act within the law.

5 An employee, with four years satisfactory service, is dismissed for reasons of unacceptable conduct. Trace the procedures which will result in the adjudication of the case by an industrial tribunal.

8 Safety administration

8.1 The Health and Safety at Work etc Act 1974	8.1.1 General duties 8.1.2 The administration of the Act 8.1.3 Statutory Notices 8.1.4 Offences

8.2 The Safety Representatives and Safety Committee Regulations 1977	8.2.1 Safety representatives 8.2.2 Safety committees

8.3 Safety policy and organisation	8.3.1 Policy statements 8.3.2 Organisation for safety 8.3.3 Safety and job descriptions 8.3.4 The safety officer

8.4 Accident and dangerous occurrence reporting and investigation	8.4.1 Reporting accidents 8.4.2 Dangerous occurrences 8.4.3 Accident investigation

Despite the dangerous conditions which must have prevailed in the construction industry since man first started to build, it was not until the Building (Safety Health and Welfare) Regulations came into effect in 1948 that legal requirements written specifically for the industry were produced. This comprehensive document covered the majority of circumstances pertinent to construction and introduced the Safety Officer into firms employing more than 50 persons.

The Factories Act 1961 consolidated those of 1937, 1948 and 1959 and introduced the Construction (General Provisions) Regulations which required companies employing in excess of 20 persons to appoint in writing a Safety Supervisor who would:

1 Advise the contractor on the observance of the statutory requirements.

2 Promote and monitor safe systems of work.

Also in 1961 the Construction (Lifting Operations) Regulations were issued and in 1966 the Construction (Working Places) and the Construction (Health and Welfare) Regulations completed the move to unity between building and civil engineering. In 1974 what was potentially the most important piece of legislation was created with the enactment of the Health and Safety at Work etc Act 1974 which provided a legislative framework to promote, stimulate and encourage high standards of health and safety in work.

An individual may find himself therefore bound to act in a safe manner under three types of law. Common law has long established a duty of care under which no one may behave in a manner that will endanger others, and parties to a contract may find that some clauses are contained which specifically relate to safety. Both of these cases, being civil, require damage to be done or loss to be experienced before a claim for compensation can be entertained. However statute law, that created by Act of Parliament, can lead to fines and imprisonment for employers creating hazards irrespective of whether an accident has happened or not.

8.1 THE HEALTH AND SAFETY AT WORK ETC ACT 1974 (H & SWA)

This Act, which received Royal Assent on the 31st July 1974, is superimposed on existing legislation and until detailed regulations and codes of practice are issued existing duties will be continued. The Act is divided into four parts:

Part 1 — covers health, safety and welfare at work.

Part 2 — establishes the Employment Medical Advisory Service.

Part 3 — enables the amendment of the Building Regulations.

Part 4 — contains a number of miscellaneous provisions.

The objectives of Part 1 are designed to ensure:

1 The maintenance and improvement of standards of health, safety and welfare of people at work.

2 The protection of the public generally from risks in connection with the activities of people at work.

3 The control of noxious and dangerous substances.

4 The control of the emission of dangerous or noxious substances.

5 The setting up of the administration required by the operation of the Act.

8.1.1 GENERAL DUTIES

In order to fulfil the main duties of ensuring the safety, health and welfare of all his employees, the employer must:

1 Provide sites and access, plant and methods of work which are safe and without risk to health.
2 Provide employees with sufficient information, training and supervision as is necessary to ensure their safety and health.
3 Provide and maintain a safe and healthy working environment.
4 Provide and maintain good welfare and hygiene facilities.
5 Ensure that the work on the site does not expose anyone nearby or any visitor to site to risk.

All of these provisions, as is the main duty, are subject to the phrase 'so far as is reasonably practicable'. This phrase has through the judicial system an interpretation which generally balances the cost of injury or disease to the worker against the cost of improvement to the employer. The following of legal precedence, codes of practice, advice of trade associations and negotiations with accredited employee representatives should ensure that agreement is reached as to what in particular circumstances is 'reasonably practicable'. In addition to these factors an employer must:
1 If five or more people are employed, provide a written statement of general policy and bring it to the attention of all employees.
2 Carry out all the requirements of the Safety Representatives and Safety Committee Regulations 1977.

Duties are also placed on employees who must:
1 Take reasonable care of their own and their colleagues health and safety.
2 Abide by all health and safety regulations, and those agreements pertinant to the site.
3 Co-operate on health and safety matters with the employer.
4 *Not* abuse any facility provided in the interest of health and/or safety.

Designers, manufacturers and suppliers of both materials and plant are also constrained by the following duties:
1 Ensure that their product (so far as is reasonably practicable – again) is safe and without risk to health.
2 See that it has been adequately tested.
3 Provide information regarding its use and detail all necessary safety precautions.

8.1.2 THE ADMINISTRATION OF THE ACT

Fig 8.1 shows the mechanisms that have been set up to administer the provisions of the Act.

The Health and Safety Commission The Commission consists of representatives of both sides of industry and local authorities and takes over the responsibility of policy development from government departments. Its purpose is to ensure that:
1 Adequate advice and information is available.
2 Research and training is carried out.
3 New regulations and codes of practice are prepared as necessary.
4 The requirements of the Secretary of State for Employment are carried out.
5 Investigations into major disasters are instituted.

The Construction Industry Advisory Committee (CIAC) has been established to advise the Commission on those matters individual to the construction industry. It comprises representatives of employers' associations and trade unions with the Chairman and secretarial services being provided by the Health and Safety Executive.

The Health and Safety Executive The Executive was established on January 1st 1975 to carry out the directions of the Commission. It has specific duties to make all

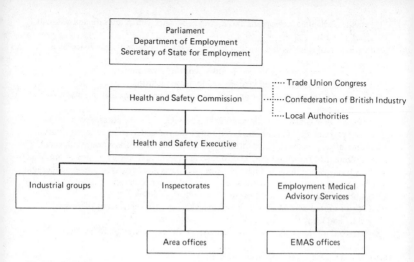

Fig 8.1 Mechanism for the administration of the Health and Safety Legislation

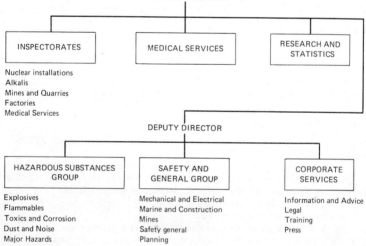

Fig 8.2 The structure of the Health and Safety Executive

necessary arrangements for the enforcement of the provisions of the Act. Its structure is illustrated in *Fig 8.2*

HM Factory Inspectorate In the construction industry the statutory provisions are enforced through the Factory Inspectorate arm of the Health and Safety Executive. Inspectors have very wide powers and may:

112

1 Enter premises.
2 Take a policeman or any other person or equipment to assist the investigation.
3 Require to be left undisturbed during the investigation.
4 Take measurements photographs and samples.
5 Dismantle, subject to test or take possession of any item.
6 Require a person to answer questions and produce relevant documents.

8.1.3 STATUTORY NOTICES

If an Inspector considers that a provision of the Act has been contravened he may serve an Improvement Notice which will require the remedy of the stated defect within a specified time period. On receipt of such a notice the offending company can:
    1 Carry out the remedial work within the time period.
or 2 Obtain withdrawal by the inspector within the time period.
or 3 Submit an application for extension within the time period.
or 4 Appeal against the terms of the notice within 21 days of it being served. The notice is then suspended until the appeal is heard.

    If the contravention is serious and there is a risk of injury a Prohibition Notice may be served. This notice also specifies details of the contravention and the time at which it will come into effect. If the commencement is deferred then it may be withdrawn by the Inspector or an appeal may be made to an Industrial Tribunal. This time however the notice stays in effect until a decision is reached. The effect of the notice is to stop the activity which has given rise to the risk until remedial action has been taken.

8.1.4 OFFENCES

Anyone who fails to carry out their duties under the Act commits an offence, irrespective of whether an accident occurs, and is thus liable to criminal prosecution. If found guilty by the courts a fine may result, which, if the contravention continues, may be reimposed on a day-to-day basis. A number of specific offences, including failure to comply with a prohibition notice are indictable and through a higher court punishable by an unlimited fine and a two year term of imprisonment.

    If it can be shown that the offence has been committed in the full knowledge, or through negligence, of any responsible individual, a managing director or site manager for example, then that individual may be fined or imprisoned.

8.2 THE SAFETY REPRESENTATIVES AND SAFETY COMMITTEES REGULATIONS 1977

These Regulations, issued under the powers enacted by the HSW Act came into operation on the 1st October 1978. They require the appointment of safety representatives, describe their functions and training and provide for the establishment of safety committees. For those parties to the NJCBI the regulations are covered by Working Rule 24 of the Working Rule Agreement and any differences in interpretation can be dealt with in accordance with the agreed procedure for the settlement of disputes.

Independent trade unions recognised by the employers, through the NJCBI for example, are given the right to elect representatives. The number of such safety representatives is not specified and is usually the subject of negotiation in which the following factors should be considered:

1 Number of operatives employed.
2 Variety of occupations and trade unions.
3 Nature of the work in respect of dangers involved.

Appointed representatives have a number of listed functions but do not accept any legal responsibility for their execution. The employer does have a responsibility to consult and co-operate with a properly appointed representative on any matter related to those functions which are listed below:

1 To investigate potential hazards, dangerous occurences and accidents at the workplace.
2 To investigate the complaints of the operatives represented related to safety, health and welfare.
3 To report, preferably in writing, on the results of any investigation or other general matters.
4 To make inspections of the workplace.
5 To represent the workforce on safety committees, and to the HSE and enforcing authority.
6 To receive information from inspectors.
7 To encourage co-operation between employer and employees on matters relating to safety, health and welfare.

In order to carry out these functions the representatives will need to carry out a number of inspections and it is recommended that details regarding the frequency, notification and relationships with the company safety officer are determined through joint negotiation. *Table 8.1* summarises different types of inspection.

Table 8.1 Types and timing of inspections undertaken by Safety Personnel

Types of inspection	*Frequency*
Regular	Three monthly
Change in working conditions	Immediately following
Accident or dangerous occurance	Immediately following
Disease	On identification
Employee complaint	On receipt
New hazard information	On receipt
Relevant documents (subject to legal and security constraints)	Continuous

The regulations provide for paid time off work for the representative to carry out his duties and to attend recognised training. The procedures for dealing with time off are regulated by a Code of Practice issued by the HSC in 1978 entitled *Time off for training of safety representatives: a code of practice.* It is recommended that agreement be reached between trade unions and the employer through a nominated, suitably trained, representative in relation to the extent of initial and further training courses.

8.2.2 SAFETY COMMITTEES

If two safety representatives feel that a safety committee will make an effective contribution to co-operation between management and labour in matters relating to safety, health and welfare, then, following negotiation, such a committee should be established within 3 months of the request. The establishment, constitution and working practices are probably best specified in a written agreement which may include:

1 Membership and Officers of the committee.
2 Functions, procedures and meetings programme.
3 Channels of communication.

The work of the committee will include:

1 The study of statistics and safety and health trends.
2 Consideration of reports submitted by management, safety representatives and enforcing authorities.
3 Monitoring of communication and training.
4 Assistance with the development of safety rules and safe systems of work.

The committee should be under the chairmanship of a senior management representative and receive secretarial support for the production and distribution of papers, agenda and minutes. Other members should include site management and functional representatives, sub-contractors, safety officers and safety representatives in addition to other nominees of the operatives to ensure a balance between management and labour representation.

There is no formal relationship between safety representatives and the safety committee, neither being nominated by the other, but in order for them to work effectively their relationship should be supportive.

8.3 SAFETY POLICY AND ORGANISATION

It is required by the Act that all firms having five or more employees must prepare a written statement of policy on health and safety at work and that arrangements are made to ensure that this policy is carried out.

8.3.1 POLICY STATEMENT

The policy statement of any firm will be determined by many factors and many variants are possible to suit individual circumstances. The following considerations will ensure that when safety policy is considered the results meet the legal requirements:

1 The awareness of legal, personal and economic responsibilities for safety.
2 The specification of responsibility for accident prevention.
3 The provision of all necessary training.
4 Liaison with external bodies.
5 The encouragement of participation.
6 The investigation, analysis and costing of accidents.
7 The consideration of safety matters at tendering and planning stages.
8 The provision of funds to implement the policy.
9 The provision and maintenance of all necessary safety equipment and protective clothing.
10 The arrangements to be made with subcontractors and others sharing the site.

Chapter 1 showed that there can be no single organisation which will suit all companies and the consideration of the safety function is no exception. There are however three guiding principles that should be followed:

1 The responsibility for safety must not be divorced from the responsibility for production.
2 However the monitoring of the manner in which those responsibilities are carried out should be so divorced.
3 This monitoring function should have representation at the highest level.

These principles are reflected in the organisation charts shown in *Fig 8.3* applicable to small, medium and large firms. The charts should be read with the job descriptions and responsibilities listed later.

Many firms cannot, considering their level of activity, satisfactorily employ a safety officer on a full time basis. Rather than share the function with some other, thus

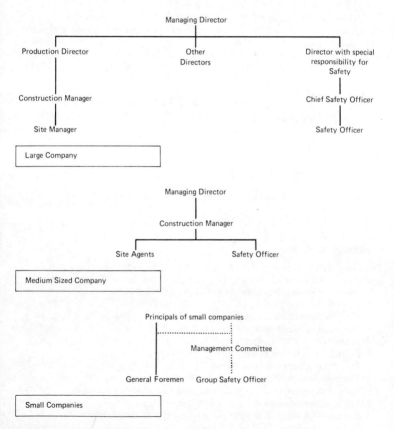

Fig 8.3 Organisation of the safety function

creating divided loyalties, it is possible for a number of small companies, through a management committee, to employ a group safety officer to monitor standards on site and give assistance and advice when required.

The organisation structure does not however ensure a safe site, in the same way that it does not ensure a productive one. The following factors will in the end determine the accident level on site:

1 The knowledge and acceptance of safety policy by all.
2 The ability to recognise potential dangers.
3 The motivation of production workers.
4 The co-ordination of work activity.
5 The effectiveness of managerial communication.
6 The availability and use of correct resources and equipment.

8.3.3 SAFETY AND JOB DESCRIPTIONS

If the responsibilities of production management necessarily include aspects of safety this must be reflected in job descriptions. The following clauses or extensions of them should be included:

Director with special responsibility for safety or Principal of a small firm. To initiate, review, implement and monitor safety, health and welfare policy in accordance with the letter and the spirit of all relevant legislation.

Construction Manager To implement company safety policy in the planning and execution of all work carried out by the company.

Site Manager To plan and maintain a site free from all foreseeable hazard, and to ensure that all instructions are precise, clear and specify all necessary precautions and to provide all necessary equipment.

Trade Foremen To incorporate safety instructions in all routine direction and to ensure that all safety equipment is used and that no unnecessary risks are taken.

All levels of management must be aware of the force of personal example particularly in respect of the wearing of safety helmets and clothing. Disciplinary procedures must be consistently operated in respect of violations of rules, instructions and agreed procedures. The absolute responsibility of superiors for the actions of their direct subordinates ensures diligence in the monitoring of performance in relation to productivity. Equally objectives must be set in the area of safety and performance reviews must give equal emphasis to the comparison of performance with the objectives set.

8.3.4 THE SAFETY OFFICER

It is important that, where possible, Safety Officers are employed on a full time basis and are trained and experienced in the construction industry. They must have detailed knowledge of all legislation and regulations and be aware of their correct interpretation. It is probable that they might aspire to membership of the construction section of the Institute of Safety Officers.

The major duties of a Safety Officer will include:

1 To advise management on all matters related to safety.
2 To carry out surveys in association with site management and safety representatives to determine the safety of operations and to make recommendations for their improvement.

SAFETY OFFICER'S REPORT	SITE DATE
ITEM	COMMENT

Plant and Equipment
   Scaffold and Means of Access
   Lifting Appliances
   Chains,Ropes and Lifting Gear
   Excavation and Demolition
   Plant and Machinery
 	
Materials Storage
 	
Documentation
   Notices
   Records
 	
Safety Equipment
 	
First Aid Facilities
 	
Health and Welfare Facilities
 	
Site Conditions	

Copies to
   Site Manager
   Construction Manager
	Signed ....
	                                  Safety Officer
	Date ....

Fig 8.4 Typical safety officer's report form

118

3 To advise management at tendering and planning stage of potential hazards and recommend appropriate precautions.
4 To carry out investigations on all accidents and dangerous occurrences.
5 To encourage and establish safety training and generate participation between operatives, management and external organisations.
6 To monitor the recording and supervise the analysis of information related to safety performance.

The formal lines of communication between the Safety Officer and site management staff must be reinforced with attitudes of mutual respect and understanding. This will be assisted by the use of standard procedures before, during and after site inspections, the results of which might well be recorded on a typical form illustrated in *Fig 8.4*.

8.4 ACCIDENT AND DANGEROUS OCCURRENCE REPORTING AND INVESTIGATION

From the 1st January 1980, the Notification of Accidents and Dangerous Occurrences Regulations 1980, repealed Section 80 of the Factories Act 1961 and installed a new procedure for the investigation, recording and reporting of accidents.

8.4.1 REPORTING ACCIDENTS

The procedure for reporting an accident varies dependent upon whether it is considered to be 'notifiable' or 'reportable'.

Notifiable accidents are those which:
1 Cause death or major injury to an employee. A major injury is specified as certain fractures, amputations, loss of sight of an eye or hospital treatment lasting more than 24 hours.
2 Occur at the workplace and cause death or major injury to a self-employed person or member of the public.
3 Include any specified dangerous occurrence even if nobody is injured.

Reportable accidents are those which result in more than three days absence from normal work.

The procedure to be followed after a notifiable accident or dangerous occurrence is:
1 Report the incident to the HSE by telephone if possible.
2 Record details of the event in the company official record system or on Form number F2509.
3 Within 7 days complete and submit Form F2508 to the HSE.

The procedure following a reportable accident is:
1 Record the details of the event in the record system or on Form F2509.
2 Complete Form B 1 76 when asked to do so by the Department of Health and Social Security.

Records, whether kept on a company system or on the official forms, must be kept for three years and must be made available on request to the HSE and Safety Representatives.

8.4.2 DANGEROUS OCCURRENCES

A full list of incidents which constitute dangerous occurrences is to be found on Form F2508 but the ones of major interest to the construction industry are:

1 The collapse or overturning of any lift, hoist, crane, excavator or mobile powered access platform, or the shedding of any load, which in all the circumstances might have caused a major injury.

2 A collapse or part collapse of any scaffold more than 12 m high which results in a substantial part of the scaffold falling or overturning.

3 The collapse or part collapse of any building, structure of falsework involving more than 10 tonnes of material.

8.4.3 ACCIDENT INVESTIGATION

In order that the requisite information can be given to the authorities a detailed investigation will be necessary. Standard procedures should be devised with the roles of site management, safety representatives and safety officer clearly defined. The investigation should ascertain:

1 The names of people involved, their occupation and injuries.
2 Names of witnesses.
3 Time, place and date of occurrence.
4 Treatment administered.
5 Events leading up to, and details of, the occurrence.
6 Details of equipment and plant involved.
7 The authorisation and instructions for the work being carried out.

Statements, photographs and diagrams will generally assist in recording the detail which will enable accurate recall of all the circumstances which might be subsequently required. In addition to fulfilling the statutory requirements the information collected should be used to compile reports and statistics to assist inter-site and inter-firm comparison and for presentation to safety committees and the company directorate.

A number of indices have been developed to provide a meaningful basis of comparison:

$$\text{Frequency Rate} = \frac{\text{Number of lost time accidents}}{\text{Total man hours worked}}$$

(use when man hours worked exceed 100 000)

$$\text{Incidence Rate} = \frac{\text{Number of reportable/notifiable accidents}}{\text{Number of employees}}$$

$$\text{Mean Duration} = \frac{\text{Number of man hours lost}}{\text{Number of lost time accidents}}$$

$$\text{Severity Rate} = \frac{\text{Man hours lost} \times 100}{\text{Total hours worked}}$$

[The term 'lost time accidents' has no single definition indeed all accidents cause some lost time however slight. The company may decide on a definition which is most meaningful which could be an accident causing loss of time beyond the day (or ½ day) in which it occurs.]

The use of these indices over a long period of time allows a company to make quantitative assessments of their safety performance and provides objective evidence of improvement or deterioration.

BIBLIOGRAPHY

The Chartered Institute of Building. Annotated Bibliographies

   38 *Site Welfare*

   74 *The Health and Safety at Work etc Act* (1974)

Croner's Reference Book for Employers. Amendment Service, Croner Publications Ltd

National Federation of Building Trade Employers, *Construction Safety Manual,*
NFBTE London

Health and Safety Commission Leaflets

   The Act Outlined. HSC2

   Advice to Employers. HSC3

   Advice to Self Employed. HSC4

   Advice to Employees. HSC5

Health and Safety Executive

   Reporting an Accident. Leaflet HSE 11

   Reporting an Accident: What you must do. Form 2508

Form 2509

The Royal Society for the Prevention of Accidents, *Safety at Work: A Guide for
Management,* IS/106 ROSPA, Purley.

EXERCISES

RECALL QUESTIONS

1 Complete the following headings from a Safety Officers Report Form

    (a) P . . . . & E . . . .   (e) F . . . . A . . F . . . .

    (b) M . . . . S . . . . . .   (f) H . . . . . & W . . . . . F . . . .

    (c) D . . . .   (g) S . . . C . . . .

    (d) S . . . . . E . . . .

2 Identify the following bodies concerned with safety

    (a) HSC;

    (b) EMAS;

    (c) HSE;

    (d) CIAC..

3 List six types of inspection which may be carried out by a Safety Representative

    (a) . . . .   (d) . . . .

    (b) . . . .   (e) . . . .

    (c) . . . .   (f) . . . .

4 Insert in the correct sentence

    Prohibition Notice

    Improvement Notice

    Notifiable Accident

    Reportable Accident

    (a) A . . . . stops an activity giving rise to risk of serious injury.

    (b) An accident which causes major injury to an operative is classified as a . . . .
. . . .

    (c) An appeal against a . . . . suspends the service of the notice.

    (d) A . . . . is notified to the HSE by the Department of
Health and Social Security.

5   Complete the following formulae

Incidence Rate =   ───────────────────

= $\dfrac{\text{Man hours lost} \times 100}{\text{Total hours worked}}$

Mean Duration = $\dfrac{\text{Number of} \ldots\ldots\ldots}{\text{Number of} \ldots\ldots\ldots}$

= $\dfrac{\text{Number of} \ldots\ldots\ldots}{\text{Total man hours worked}}$

QUESTIONS REQUIRING SHORT ESSAY ANSWERS (15–20 minutes)

1   Describe the powers of a Factory Inspector in relation to construction site inspection.

2   You have been appointed the Chairman of a newly established Safety Committee. Prepare an agenda for the first meeting and make brief notes from which to give a welcoming address to the members.

3   Identify the potential safety hazards which exist within your job and describe the action you take to overcome them.

4   List the functions of a Safety Representative as contained in the relevant guidance notes.

5   Draft a suggested form to be used by a Safety Representative to report to site management potentially unsafe or unhealthy conditions at work.

QUESTIONS REQUIRING LONGER ESSAY ANSWERS (30–45 minutes)

1   Examine the project shown on drawing no . . . . and identify all potential hazards in respect of:
(a) Materials storage and movement.
(b) Access for employees and visitors.
(c) The use of mechanical plant.

2   Prepare a report for site management designed to minimise all the hazards listed in answer 1.

3   'There is no such thing as an accident'. Comment upon this statement in relation to building work.

4   Obtain a copy of your company's Policy Statement on Safety, Health and Welfare and analyse its contents in relation to the general duties of employers as set out in section 2 of the H & SW A 1974.

5   Prepare a series of procedures designed to meet fully the requirements of the Notification of Accidents and Dangerous Occurrences Regulations 1980.

9 Education and training

9.1 The organisation of training	9.1.1 Manpower Services Commission and Training Services Agency 9.1.2 The Construction Industry Training Board 9.1.3 Building Advisory Service 9.1.4 Other training organisations
9.2 Further education	9.2.1 The Department of Education and Science 9.2.2 Local Education Authorities 9.2.3 Colleges of Further Education 9.2.4 The Polytechnics 9.2.5 The Universities 9.2.6 The Council for National Academic Awards 9.2.7 The Technician Education Council 9.2.8 The City and Guilds of London Institute
9.3 In-company training	9.3.1 The cost of training 9.3.2 Systematic training 9.3.3 Organisation of training 9.3.4 General training 9.3.5 Apprentice training in the building industry
9.4 Professional and Technician Institutions	9.4.1 The Royal Institute of British Architects 9.4.2 The Chartered Institute of Building 9.4.3 The Chartered Institute of Building Services 9.4.4 The Institution of Structural Engineers 9.4.5 The Royal Institution of Chartered Surveyors

Many suggestions have been made as to the difference between education and training. Generally speaking training is taken to be that activity which attempts to improve a person's skill in a particular task as opposed to education which deals with general personal development. From this distinction it follows that the objectives of training are more immediate and measurable, the time span is shorter and a cost benefit analysis more readily available and meaningful. However it must be recognised that the two are difficult to identify within any programme.

Education and training within the building industry can be considered in four sections:

1 The Organisation of Training.
2 Further Education.
3 In-Company Training.
4 Professional and Technician Bodies.

9.1 THE ORGANISATION OF TRAINING

At a national level, the organisation of training has, since 1973, been the responsibility of the Manpower Services Commission. The Employment and Training Act of that year and another with the same name in 1981 ammended the Industrial Training Act 1964 which originally established the Industrial Training Boards. The current organisation is illustrated in *Fig 9.1*.

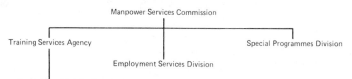

Fig 9.1 The organisation of the Manpower Services Commission

9.1.1 MANPOWER SERVICES COMMISSION (MSC) AND TRAINING SERVICES AGENCY (TSA)

The MSC comprises a full time Chairman and nine part-time members drawn from industry, trade unions, local authorities and education. Its main duties are to assist people in the selection, training and obtaining employment and to assist employers by ensuring the availability of skills in the workforce.

The TSA is responsible for the operation of the Training Opportunities Scheme (TOPS) which:

1 Assists in the acquisition and upgrading of skills.
2 Offers special training for the disabled.
3 Retrains workers whose skills are redundant.

Training allowances are paid to those who attend TOPS courses at:

1 Colleges of Further Education.
2 Private Colleges.
3 Residential Training Centres for the Disabled.
4 Government Skillcentres.

124

9.1.2 THE CONSTRUCTION INDUSTRY TRAINING BOARD (CITB)

The Industrial Training Act 1964, as amended by the Employment and Training Acts of 1973 and 1981, was established to improve training in quality and quantity and distribute the cost more equitably amongst all employers. The mechanism by which this was to be achieved was by the establishment of twenty-five Industrial Training Boards. This number was reduced in 1982 to eight of which the Construction Industry Training Board was one. An ITB is empowered to:

1 Establish courses of training.
2 Approve courses established by others.
3 Investigate and publish training recommendations.
4 Instigate research into training.
5 Provide advice and guidance where required.

Basically Boards are financed on a levy grant system whereby employers pay a levy on their payroll and then receive grants for approved training. In 1973 the Government of the day undertook to pay the operating costs of the Board, which in 1981 amounted to £4½m but this was withdrawn in March 1982 and as the levy cannot be raised, further changes will have to be made.

To date approved training for which grants are payable have included:

1 Attendance at specified approved day and block release courses of further education and training.
2 The provision of places to students on industrial training as part of a sandwich course of further education.
3 Attendance on approved private courses of training.
4 Attendance on approved in-company training.
5 Group training schemes offered on a non-profit making basis.

In addition to these the Board operate four residential training centres which offer courses in craft skills and supervision.

9.1.3 BUILDING ADVISORY SERVICE (BAS)

BAS is the training arm of NFBTE. In addition to operating advisory services and training in safety and environmental control, the following courses are offered:

General management.
Financial and insurance.
Contractual.
Site management and industrial relations.

Typically a course may run for 2—3 days on both residential or non-residential bases and be repeated at various venues around the country. Fees are charged but NFBTE members enjoy preferential terms.

9.1.4 OTHER TRAINING ORGANISATIONS

Training is also available from the following sources:

1 Trade Associations such as the Cement and Concrete Association and the Timber Research and Development Association.
2 College of Further Education, Polytechnics and Universities offer short courses in practical areas.
3 Professional and Technician Bodies arrange local and regional events from single evenings to week long conferences.
4 Manufacturers offer training facilities in relation to new products.

5 Trade unions have a developing interest in training.
6 Plant and equipment manufacturers operate training schools on a commercial basis.
7 Private training companies operate courses on a wide range of subjects at a number of venues.

9.2 FURTHER EDUCATION

Further education, by definition, includes all post secondary education excluding universities, and therefore covers the whole spectrum of courses from craft studies to degrees. When a person leaves school the decisions taken in regard to further education will depend upon four factors:

1 Career aspirations.
2 Education patterns relative to that career.
3 Results in examinations.
4 Availability of opportunity.

Within the construction industry four broad overlapping categories of career potential may be identified:

1 *Technologists* — Generally fully qualified professionals in design, planning, technology and management who will qualify through reading for a degree or by taking the examinations offered by the professional institutions.
2 *Technicians* — Generally carry out a wide range of responsible jobs between the technologist and the operative involving mathematical, scientific and technical skills. They will qualify through certificate and diploma examinations or again may take examinations offered by technician bodies.
3 *Craft operatives* — Generally are skilled workers who have undergone a formal apprenticeship combining on-the-job training with part time education leading to an award of the City and Guids of London Institute.
4 *Operatives* — Generally require less skill and knowledge than the craft operative which is developed mainly through experience, although increasingly a variety of ad-hoc courses are organised, often through the Construction Industry Training Board.

Knowledge of further and higher education requires the examination of three categories of organisation:

1 The authorities controlling the provision of education.
2 The institutions offering education.
3 The bodies controlling the awards.

9.2.1 THE DEPARTMENT OF EDUCATION AND SCIENCE (DES)

Since 1964 the Department of Education and Science, headed by the Secretary of State for Education, has been responsible for the whole education system in England and post-secondary education in Wales. The Scottish non-university sector is under the control of the Secretary of State for Scotland and Northern Ireland has its own Department of Education.

The Department originates the general policy for education and controls the broad allocation of funds which central government provide for local authority spending on education.

The Department:
1 Sets minimum standards.

2 Controls spending on buildings.
3 Controls teacher training.
4 Controls funds from central government.
5 Settles disputes between parents and local education authorities.

It does not:
1 Administer schools of colleges.
2 Set the curriculum.
3 Prescribe textbooks.
4 Engage teachers.

9.2.2 LOCAL EDUCATION AUTHORITIES (LEA)

Local Education Authorities are the local government bodies responsible for the provision of education in schools, colleges and polytechnics, but not in universities. There are 122 such authorities in Great Britain

England and Wales	105
Scotland (Education Authorities)	12
Northern Ireland	5

The Authorities operate through Education Committees which include in their membership a majority of elected councillors. Although influenced by national policy, through statutory direction, orders and financial control, the Committees exercise a degree of autonomy which account for the many regional variations in education provision. LEAs:
1 Build and own schools and colleges.
2 Maintain, supply and inspect establishments.
3 Train, appoint and dismiss teachers and other staff.
4 Enforce school attendance.
5 Award grants to some students in further education.

9.2.3 COLLEGES OF FURTHER EDUCATION

As further education covers all post secondary education ranging from craft study to degrees the titles of colleges and their organisation vary from Authority to Authority. At most, a wide range of qualifications both national and college is available which can be gained through a variety of study methods:
1 *Full time* — Attendance at college for the full course duration.
2 *Sandwich* — College based students spend a specified period gaining experience in industry.
3 *Block Release* — Students based in industry spend periods of full time attendance in college.
4 *Part time* — Students based in industry spend a period of every week studying in college.

9.2.4 THE POLYTECHNICS

In 1966 it was decided to centralise advanced courses within institutions which were to have greater, but not complete, autonomy from LEAs and since that date thirty such institutions have been created in England and Wales. They provide a variety of technical, professional and degree courses in all modes of attendance with entry qualifications set at A level or equivalent. They also enjoy close links with industry

127

and are required to consider the specific needs of that industry when courses are designed.

9.2.5 UNIVERSITIES

Universities are self-governing bodies which, although financed largely by central government through the University Grants Committee, do not come under the control of either central or local government. They hold their rights through the grant of a Royal Charter from the Privy Council. The Universities themselves decide, within the financial constraints, which degrees to award, which students to enrol and which staff to appoint.

The awards of universities are varied but those applicable to the building industry are in the main:

1 Batchelor of Science (BSc) awarded after three or four years of full time or sandwich study. The entry qualifications for such a course are generally five GCE subjects (Grade C or better) (or equivalent) of which two must be at A level. Certain technical qualifications are also usually acceptable.

2 Master of Science (MSc) awarded after a variable period of formal study and examinations or research culminating in the submission of a thesis. Attendance may be full or part time.

3 Doctor of Philosophy (PhD) awarded after a longer period of research and is almost invariably reserved for the submission of an original thesis upon which an oral examination is conducted.

Universities also award a number of post graduate qualifications which are individual to that institution.

9.2.6 THE COUNCIL FOR NATIONAL ACADEMIC AWARDS (CNAA)

The Council was founded by Royal Charter in 1964 to award degrees and other

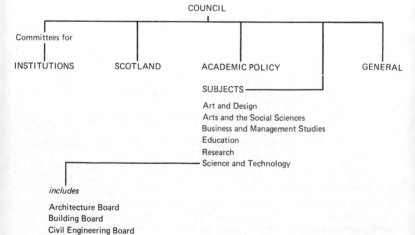

Fig 9.2 The Committee Structure of the Council for National Academic Awards

academic distinctions to students who complete approved courses or carry out research in institutions other than universities. Its work includes the approval of courses submitted by institutions and then, through formal visits and examination moderation, to maintain and co-ordinate national standards. The awards, methods of study and entry qualifications are generally the same as those described for universities. About one-third of all degree students are currently attending CNAA approved courses, mainly at Polytechnics.

The committee structure of the Council for National Academic Awards is shown in *Fig 9.2.*

9.2.7 THE TECHNICIAN EDUCATION COUNCIL

The Technician Education Council was established in 1973 to plan, administer and keep under review a unified national system of technical education. In building the changeover is virtually complete and the vast majority of students commencing their studies in 1982 will eventually obtain an award of this body. Institutions may submit schemes based upon national guidelines which are channeled for approval through Programme Committees, those for construction being:

B2 Architecture, Building and Quantity Surveying.

B3 Building Engineering Services.

B4 Civil and Structural Engineering.

B5 Cartography, Planning and Land use.

The programmes are validated by TEC in terms of content, assessment and comparability of standards and lead to the awards of:

Certificate.

Diploma.

Higher Certificate.

Higher Diploma.

Each award is based upon the acquisition of a standard number of 60-hour subject units at various levels and is therefore independent of the mode of study. Differing combinations of units may also be taken thus providing a range of subjects allied to the student's current or intended career. Entry onto a certificate or diploma programme requires a five year secondary education with some level of attainment specified in some circumstances. Direct entry onto higher award programmes may be possible necessitating bridging studies.

9.2.8 THE CITY AND GUILDS OF LONDON INSTITUTE (CGLI)

The City and Guilds of London Institute has a long history and still remains the largest technical examining body in Britain. It produces schemes of education, sets examinations and establishes national standards. Courses leading to CGLI qualifications are offered by colleges of further education and also by Government Skillcentres and are mostly part time day release although some are offered on block release and evening only basis. Although the CGLI offered both technician and craft qualification in the construction industry the former have in the main been replaced by the new TEC courses. Craft certificates and advanced craft certificates are offered in the building crafts.

Fig 9.3 attempts to summarise the relationships between the authorities, the institutions of education and the awards they offer.

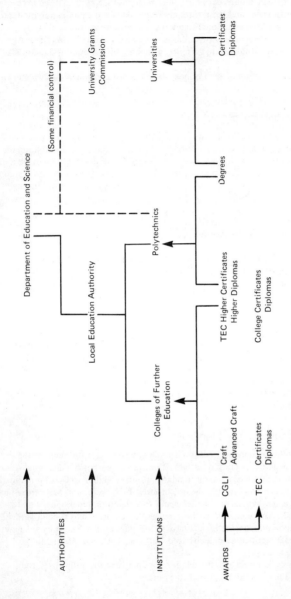

Fig 9.3 The authorities and institutions of further education

9.3 IN-COMPANY TRAINING

Traditionally all training undertaken by workers was in-company training and generally took place at the work face by firstly observing a skilled operative and then by attempting the task. Whilst this method has inherent disadvantages it must be recognised that skills, however learned have eventually to be transferred to the task and so in-company training is inevitable.

The situations where on the job training is ineffective are those in which skills are to be learned which:

1 Cannot be learned effectively or safely whilst in the process of doing the job.
2 Demand the skills of highly qualified instructors and/or expensive training equipment.
3 The need to involve a variety of departments.

In these cases off-the-job training is necessary and the difficulties of releasing the person for training, setting up the learning experience and transferring the learning must be faced.

9.3.1 THE COST OF TRAINING

A cursory examination of the cost of training would lead to the consideration of such factors as the direct cost of the course and the loss of production; but such a study should also consider the following:

1 The cost of not learning as reflected in a permanently low production rate.
2 The cost of learning by doing as reflected by:
   Wages paid for ineffective performance.
   High materials waste.
   Unacceptable work remedied or rejected.
   Damage to machinery.
   Increase in accidents.
   High labour turnover.

Fig 9.4 The training balance

3 The costs of training can be divided into the direct costs, such as the training officer and the operatives wages, and the indirect costs such as overheads and preparation time.

Systematic training will lead to reductions in:

1 Learning time.
2 Production stoppages.
3 Waste of material.
4 Leaving rate.
5 Supervision time.

When considering the cost of training it is necessary therefore to prepare a balance sheet in which the credits and debits are balanced as shown in *Fig 9.4.*

9.3.2 SYSTEMATIC TRAINING

Chapter 3 established the requirement for a manpower plan and systematic training will ensure that an effective workforce is available to help the company reach its objectives. Systematic training can be divided into 4 stages:
  1 Identification of the need.
  2 Planning to meet the need.
  3 Carrying out the training.
  4 Evaluating the effectiveness.

Identification of the need This may be established by answering the questions:
  1 Who needs to be trained?
  2 What difficulties will they face?
  3 What should they know?
  4 What do they know?
The difference between the last two answers is called the ability gap and the initial step is, through correct selection and job design, to reduce it to a minimum. The sum of the ability gaps will identify the need.

Planning to meet the need The preparation of a training programme commences with a consideration of all possible approaches to the problem and the selection of the least cost solution which is consistent with company policy. Programes will include:

  The skills and knowledge required.
  The location and duration of the sessions.
  The required resources.

If a purpose made training programme is required then the stages identified in *Fig 9.5* must be considered.

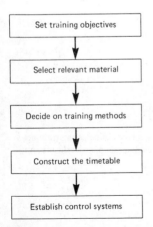

Fig 9.5 Establishing a training programme

132

Carrying out the training The programme is then installed and adjustments are made as necessary in relation to individual progress. Records are maintained to monitor individual performance which will provide feedback for evaluation purposes.

Evaluating effectiveness Reviews, both during and after the training programme, will allow trainers, trainees and line management to learn valuable lessons which can lead to improvements in the training programme. The initial objective will allow determination of effectiveness.

If the organisation is considering the use of external training courses five basic ground rules will ensure that the cost of the course will be worthwhile:
1 Set clear objectives.
2 Brief the employee properly.
3 Examine the course carefully.
4 Look for practical courses with direct, measurable benefit.
5 Ensure that the learning can be immediately implemented.

9.3.3 ORGANISATION FOR TRAINING

In order that training should be successful three organisational factors will require consideration:
1 Management must accept responsibility for training.
2 The training function must be efficiently organised.
3 The role of the training officer must be carefully defined.

Management's responsibility Management's training responsibilities for line operatives may be delegated but that for his own staff cannot. Every manager must therefore possess the necessary expertise to carry out this training role. Higher management retain responsibility for formulating company training policy and for general management training and development. The training policy will establish:
1 The extent of training.
2 The areas of priority.
3 The available resources.
Middle management are responsible for the implementation of this policy within their own department and, whilst they will become involved in coaching and on-the-job training, other aspects such as the analysis of training requirements may be delegated to training specialists with the co-operation of line management.

The organisation of the training function Fig 9.6 indicates three possible methods of organising the training function. If method 'a' is adopted and the training organised within the personnel function then its success will depend upon the attitudes of the manager of that function. It is however advantageous as training is integral with many of the other activities carried out by that department, such as recruitment and promotion.

The second case establishes a direct link between the managing director and the training function and may occur even in an organisation which possesses a personnel department. The training manager, having a direct relationship with the managing director, undoubtedly carries considerable informal authority, but as the role of top management invariably involves much absence from the office some delays might well occur in decision making.

Method 'c' can only work successfully where the training officer is responsible for one type of training only, apprentice training for example.

Like any other functional responsibility the role of training might be included as

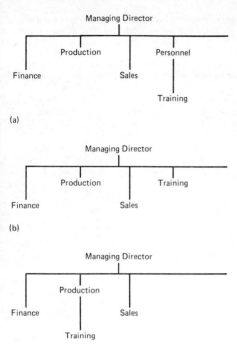

(a)

(b)

(c)

Fig 9.6 The organisation of the training function

part of a wider job description and it is not essential for a training specialist to be identified within the organisation. If such roles are separately defined however they may carry a variety of titles which include:

1 *Training Director* — is a director with a full-time responsibility for training in a very large company.
2 *Group Training Manager* — is the training manager for a group of companies which might be members of a parent group or, as is common in the construction industry, where a number of smaller companies form a training group in order to employ a specialist.
3 *Company Training Officer* — is responsible for all types of training and with assistance manages the training needs of companies employing in excess of 500 employees.
4 *Specialist Training Officer* — is required where there are large numbers of individuals with similar training needs.

The role of the training officer The role of the training officer will vary dependent upon:

1 The status and importance of the function.
2 The nature of the demand for training.
3 The calibre of the training officer and his position within the hierarchy.
4 The perceived effectiveness of past performance.

134

The responsibilities however will probably include:
1 The analysis of jobs and the identification of the training need.
2 The design and development of training programmes.
3 The management of training resources.
4 The liaison between the company and all relevant external bodies.
5 The acquisition of knowledge regarding legislative requirements, learning methods, training equipment etc.
6 The maintenance of the company training records.

9.3.4 GENERAL TRAINING

In addition to specific skill training there are a number of general training areas which must be considered an essential part of manpower development, although immediate results are not always quantifiable in the short term. Such training includes:
1 Induction training.
2 Safety training.
3 Mid-career training and retraining.
4 Supervisory training and management development.

Induction training Induction training is required by all new entrants to a company or by those people whose jobs have dramatically changed. The aims of such a course are to welcome new employees, give certain basic information about the company and explain how each individual contributes to the attainment of company objectives. The following lists the points which may require coverage:
1 General terms and conditions of employment.
2 Welfare, medical and social facilities.
3 The organisation, the market and the future.
4 The geography of the workplace.
5 Essential safety features.
6 Arrangements for training.
7 Pay arrangements and reward systems.
8 Trade union activity and grievance procedures.

Safety training Chapter 8 established the legal requirement for safety training which is categorised as follows:
1 The legal responsibilities of the company.
2 The company organisation and procedures for safety.
3 The roles and responsibilities of management, safety representatives and safety specialists.
The following is a list of sources of safety training courses:
1 The company.
2 The Construction Industry Training Board.
3 Building Advisory Service.
4 Trade Union Congress.
5 Private training companies.
Dependent upon their objectives the coverage of such courses will include a selection from the following:
1 The legal requirements.
2 Roles and functions of safety representatives and committeees.
3 Hazards at work.
4 Investigations and safety audits.
5 Joint negotiating procedures.

If it is considered necessary for a building company to use an external course, care must be taken in the selection. This will ensure that the participants will receive training relevant to construction sites where the hazards are specific and individual.

Although most courses impart knowledge and skills, safety consciousness has a psychological element which cannot be taught by traditional methods. The successful course will therefore attempt to alter attitudes towards safety and this will require very careful organisation.

Mid-career training and retraining In the past a seven-year apprenticeship period taught a person all the skills he would require in his working life. The pace of modern technology has long since rendered this situation obsolete and training for change is now a feature of most jobs. For some this will mean the acquisition of new skills and techniques to improve productivity and efficiency within the same job. For others it will mean that skills will become obsolete and retraining in another area will be necessary to prolong the working life of an individual.

These factors apply equally to production workers, management and the professions. Many of the latter are undertaking investigations into the needs of society and the resulting activity is being called Continuing Professional Development.

Supervisory and management development Supervisors play the important linking role between management and production workers and the conditions will vary from industry to industry. It is unlikely therefore that a stereotyped course will meet all the requirements of a given situation and much will depend upon the relationships between the supervisor and his superior. Supervisors and management within the hierarchy will have two responsibilities with regard to development:

1 To develop his subordinate.
2 To develop himself.

and to this end the following on-the-job techniques may be used to supplement the more formal courses of skill acquisition:

1 *Coaching* — on a one to one basis helps people to help themselves.
2 *Planned delegation* — the individual is constantly stretched thus building confidence and ability and reducing the gaps between levels within the hierarchy.
3 *Problem analysis* — solutions are sought to real problems within a realistic timetable, they may then be analysed and a decision made.

There are more formal mechanisms of management development such as job rotation and the appointment of trainees as personal assistants. It must however be recognised that the whole of a managers job is a learning situation and the allocation of time to consider these experiences and analyse the long term consequences of past decisions will play a part in future development.

9.3.5 APPRENTICE TRAINING IN THE BUILDING INDUSTRY

The training of apprentices in the building crafts is, for those working under the auspices of the NJCBI, the subject of a service agreement. Under this agreement the employer undertakes to ensure that off-the-job training and further education is supplemented by appropriate work experiences and the apprentice agrees to apply himself to the acquisition of skill and knowledge.

The National Joint Training Scheme is operated by the National Joint Training Commission through Regional and Local Joint Committees whose responsibilities include:

1 The issue of certificates and maintenance of a register of all persons completing training.

2 Liaison with all external bodies concerned with training.

3 The review of facilities provided for the training of craft apprentices.

The apprenticeship period for the crafts of carpentry and joinery, brickwork, painting and decorating and plastering is three years and that for plumbing and heating and ventilating, four years. Further education and training on CITB grant aided courses is offered through approved colleges of further education. Two schemes are available, the first of which prepares the apprentice for the appropriate craft certificate of CGLI in 2 years on a day release basis. The second scheme offered generally by the same institutions under the aegis of the CITB provides full time training for the first 26 weeks of the apprenticeship reinforced by 8 weeks of full time attendance in the second year. During the third year of the apprenticeship day release study prepares some trainees for the appropriate advanced craft certificate.

Where the full time training scheme fails to attract sufficient numbers of apprentices already in employment, the CITB may select a number of trainees who, after passing a selection test, receive a grant paid by the MSC. By the end of the 6-month full time course it is hoped to find them permanent apprenticeships.

9.4 PROFESSIONAL AND TECHNICIAN INSTITUTIONS

The term 'profession' was originally applied to law, medicine and divinity which provided the only means of earning a living which did not involve commerce or manual work. The expansion of knowledge following the industrialisation of society created new areas of specialism which led to new fully recognised professions.

The definition of the word profession is not precise and therefore considerable debate and discussion ensues the allocation of titles such as 'professional institute' and 'technician body'. The following have been identified as the characteristics of a profession:

1 It involves a skill based on a body of knowledge.

2 The skill results from education and training.

3 It requires that practitioners first demonstrate their competence.

4 A code of conduct maintains integrity.

5 It performs a recognised service for the public good.

6 It has a formal organisation.

One of the objective differentiations between institutions is the holding of a Royal Charter granted by the Privy Council. Those within the building industry are:

The Royal Institution of British Architects	RIBA
The Chartered Institute of Building	CIOB
The Chartered Institute of Building Services	CIBS
The Institution of Structural Engineers	IStructE
The Royal Institution of Chartered Surveyors	RICS

The following table lists other major institutions of the industry linked with the area of work of their members

Architects	The Architectural Association	AA
	The Faculty of Architects and Surveyors	FAS
	The Incorporated Association of Architects and Surveyors	IAAS
Architectural Technician	The Society of Architectural and Associated Technicians	SAAT

137

Building Control Officers	The Institute of Building Control Officers IBCO
Clerks of Works	The Institute of Clerks of Works ICW
Quantity Surveyors	The Institute of Quantity Surveyors IQS
Quantity Surveying Technicians	The Society of Surveying Technicians SST
Surveyors	The Construction Surveyors Institute CSI
	The Faculty of Architects and Surveyors FAS
	The Incorporated Association of Architects and Surveyors IAAS

The position regarding these institutions is never static and new bodies are created
by amalgamations, mergers and new formations. At the present time CSI and FAS
have formed a federation which will result in a merger and the IQS and RICS
will shortly merge.

9.4.1 THE ROYAL INSTITUTE OF BRITISH ARCHITECTS

The Royal Institute of British Architects was originally founded in 1834 and granted
its Royal Charter three years later. The purposes of the Institute were stated to be
'The general advancement of civil architecture and the promotion of the acquirement
of knowledge of the various arts and sciences connected therewith.'

There are two classes of membership, a non-corporate class open to bona-fide
students of architecture and the corporate class who may style themselves RIBA.
Membership of this class is gained by the passing of the RIBA's own examinations or by
gaining exemption from them and undergoing two years of acceptable training.
Members are simultaneously registered with the Architects Registration Council of the
United Kingdom (ARCUK) in order that the style 'Architect' may be used.

Most students of architecture attend courses of full time study offered at universities
and polytechnics lasting five years and resulting in the award of a degree or diploma.
The entry qualifications for such a course are five GCE subjects (grade C or better)
two of which must be at A level. Similar equivalent qualifications and certain
technician awards are also acceptable.

9.4.2 THE CHARTERED INSTITUTE OF BUILDING

The Chartered Institute of Building, whilst tracing its origin to 1834, is the latest to
receive the Royal Charter which was granted in 1980. The objects of the Institute are:
  1 The promotion for the public benefit of the science and practice of building.
  2 The advancement of public education in the said science and practice including
    all necessary research and the publication of the results of that research.

There are six classes of membership but only two, Fellows and Members, are allowed
to use designatory letters and style themselves 'Chartered Builder'. For each class of
membership there is a prescribed education and experience requirement; the former
being met by the passing of the Institute's examinations or by the possession of the
appropriate national award.

9.4.3 THE CHARTERED INSTITUTE OF BUILDING SERVICES

The Chartered Institute of Building Services, founded in 1897 as the Institution of
Heating and Ventilating Engineers, now undertake activities embracing the whole field
of building services engineering being concerned with the human, scientific and

technical aspects of the design, construction, operation and maintenance of all the built environment with the exception of the structure.

The corporate member classes Fellow (FCIBS) and Member (MCIBS), recognised as professional engineers, are open to those who possess a suitable degree or other approved qualification or who have passed the Institutes own examinations. Approved practical experience is also required. There are four non-corporate classes of membership one of which, the Associate, may register with the Engineers Registration Board as Technician Engineers and therefore use the designatory letters T.Eng (CEI).

9.4.4 THE INSTITUTION OF STRUCTURAL ENGINEERS

The Institution of Structural Engineers was founded in 1908 as the Concrete Institution, renamed in 1923 and received the Royal Charter in 1934. The Institution was a founder member of the Council of Engineering Institutions and corporate members not only use the designatory letters F or MIStrucE but also CEng, Chartered Engineer.

Examinations of the Institution comprise; the graduateship part 1 and part 2, from which a degree obtains exemption, an associate membership examination for technician engineers who attain Associate Membership and are registered TEng (CEI), and a membership part 3 examination which the majority of prospective members must take.

9.4.5 THE ROYAL INSTITUTION OF CHARTERED SURVEYORS

The Royal Institution of Chartered Surveyors was incorporated by Royal Charter in 1881, 13 years after its formation. Corporate members, Fellows (FRICS) and Professional Associates (ARICS), have qualified through practical experience and university level education and are able to use the style Chartered Surveyor. They are however each specialised in one of a variety of areas which include quantity surveyors and building surveyors.

The Institute's own examinations are conducted in three stages each lasting about two years if studied on a day release basis. The entry qualifications are set at either four GCE subjects with three at A level or five subjects with two at A level. As usual alternative equivalent qualifications are acceptable. Holders of certain degrees and diplomas may receive partial or complete exemption but in the latter case all candidates must take the Graduate Entry Examination. Once again a period of approved practical experience is required before a person is admitted to the appropriate class of membership.

BIBLIOGRAPHY

The Chartered Institute of Building. *Annotated Bibliographies. No 41 Training for Site Management,* CIOB Ascot.
The Chartered Institute of Building *List of Building Courses 1981—83,* CIOB Ascot (1981).
The Construction Industry Training Board, *Enquiry into Management Technician Roles in the Construction Industry,* CITB London, (1970).
Council for National Academic Awards (1981), *Annual Report 1980,* CNAA London
Directory of Technical and Further Education (1980), 18th ed Goodwin London (1980).

Kenney, J. and Donnelly, E.L. *Manpower Training and Development,* Harrap, London (1972)

Priestly, B. *British Qualifications,* 12th ed, Kegan Page, London (1981)

EXERCISES

RECALL QUESTIONS

1 Identify the names of the following organisations concerned with education and
   training

   BAS . . . .   LEA . . . .

   CGLI . . . .   MSC . . . .

   CITB . . . .   TEC . . . .

   CNAA . . . .   TSA . . . .

2 Complete the stages in the preparation of a training programme

   S . . t . . . .   o . . . .

   S . . . . . r . . . . m . . . .

   D . . . . . o . t . . . .   m . . . . .

   C . . . . . . .t . . t . . . .

   E . . . .   c . . . . . . s . . . . . .

3 Draw three organisation charts to indicate possible relationships between the
   productions and training functions.

4 Identify the four stages of systematic training

   (a);

   (b);

   (c);

   (d).

5 Name the following Professional and Technician Institutions

   CIBS . . . .   RIBA . . . .

   CIOB . . . .   RICS . . . .

   IAAS . . . .   SAAT . . . .

   IStructE . . . .   SST . . . .

QUESTIONS REQUIRING SHORT ESSAY ANSWERS (15–20 minutes)

1 Prepare a list of all the courses offered by your nearest college of further education
   which are of interest to the construction industry.

2 Differentiate between technologists, technicians, craft operatives and operatives
   working in the building industry.

3 Briefly describe the work of the CITB.

4 List all sources of external education and training courses available to a building
   company.

5 Compare block release and day release as methods of providing initial training for
   construction craft apprentices.

1 Differentiate between the work and organisation of colleges of further eduction, polytechnics and universities.

2 'Sending people away on courses is a waste of valuable time' Discuss.

3 Analyse in full the financial costs and benefits of an in-company training programme.

4 Prepare a syllabus and timetable for an induction course offered annually to new craft apprentices within a building company.

5 Describe in detail the role of a Company Training Officer.

List of Abbreviations

AA	Architectural Association.
ACAS	Advisory Conciliation and Arbitration Service.
ARCUK	Architects Registration Council of the United Kingdom.
BAS	Building Advisory Service.
BATJIC	Building and Allied Trades Joint Industrial Council.
BCEJB	Building and Civil Engineering Joint Board.
CAC	Central Arbitration Committee.
CBI	Confederation of British Industry.
CECCB	Civil Engineering Construction Conciliation Board.
CGLI	City and Guilds of London Institute.
CIAC	Construction Industry Advisory Committee.
CIBS	Chartered Institute of Building Services.
CIOB	Chartered Institute of Building.
CITB	Construction Industry Training Board.
CNAA	Council for National Academic Awards.
CoP	Code of Practice.
CSI	Construction Surveyors Institute.
DES	Department of Education and Science.
EA	Employment Act 1980.
EAT	Employment Appeal Tribunal.
EETPU	Electrical, Electronic, Telecommunication and Plumbing Union.
EMAS	Employment Medical Advisory Service.
EPA	Employment Protection Act 1975.
FAS	Faculty of Architects and Surveyors.
FASS	Federation of Specialist Subcontractors.
FCEC	Federation of Civil Engineering Contractors.
FMB	Federation of Master Builders.
FTAT	Furniture, Timber and Allied Trades Union.
GCE	General Certificate of Education.
GMWU	General and Municipal Workers Union.
HSC	Health and Safety Commission.
HSE	Health and Safety Executive.
HSWA	Health and Safety at Work etc. Act 1974.
IAAS	Incorporated Association of Architects and Surveyors.
IBCO	Institute of Building Control Officers.
ICW	Institute of Clerks of Works.
IStructE	Institute of Structural Engineers.
IT	Industrial Tribunal.

ITB	Industrial Training Board.
JCT	Joint Council Tribunal.
LEA	Local Education Authority.
MSC	Manpower Services Commission
NFBTE	National Federation of Building Trade Employers.
NFRC	National Federation of Roofing Contractors.
NJCBI	National Joint Council for the Building Industry.
NWR	National Working Rule.
RIBA	Royal Institute of British Architects.
RICS	Royal Institution of Chartered Surveyors.
SAAT	Society of Architectural and Associated Technicians.
SBEF	Scottish Builders Employers Federation.
SST	Society of Surveying Technicians.
TEC	Technician Education Council.
TGWU	Transport and General Workers Union.
TOPS	Training Opportunities Scheme.
TSA	Training Services Agency.
TUC	Trades Union Congress.
TULRA	Trades Union and Labour Relations Act 1974.
UCATT	Union of Construction, Allied Trades and Technicians.
WRA	Working Rule Agreement.

Index